THE DECISION TRAP

Genetic Education and
Its Social Consequences

SILJA SAMERSKI

with a preface by
Barbara Katz Rothman

imprint-academic.com

Copyright © Silja Samerski, 2015

The moral rights of the author have been asserted.
No part of this publication may be reproduced in any form
without permission, except for the quotation of brief passages
in criticism and discussion.

Published in the UK by
Imprint Academic, PO Box 200, Exeter EX5 5YX, UK

Distributed in the USA by
Ingram Book Company,
One Ingram Blvd., La Vergne, TN 37086, USA

ISBN 9781845407766

A CIP catalogue record for this book is available from the
British Library and US Library of Congress

English translation by Nancy Joyce

Contents

Acknowledgment	v
Preface to the English Edition by Barbara Katz Rothman	vii
Preface to the German Edition	xi
1. Introduction: Gene as the Basis for Decision Making?	1
Distancing as a research approach	5
2. Genetic Education	8
2.1. *The Gene*	8
2.2. *Educational Campaigns*	17
2.2.1. *Illiterate citizens? A Bremen congress*	17
2.2.2. *The genetic literacy campaign*	21
2.2.3. *Genetic counselling*	27
2.3. *On the History of Genetic Counselling:*	
Genetics as the Foundation of Sociopolitics	31
2.3.1. *The scientific management of hereditary dispositions*	31
2.3.2. *More effective than coercion:*	
Education and responsibility	33
2.3.3. *A new goal: The informed decision*	37
3. "Informed Choice": How Genetic Counsellors Empower their Clients to Attain Self-Determination	47
3.1. *The Initial Transformation of the Person:*	
The Client as a Gene Carrier	49
3.1.1. *The genetic person*	49
3.1.2. *The incomprehensible self*	51
3.1.3. *Things in the body*	56
3.1.4. *Hidden causes*	65
3.1.5. *Meaningful information*	68
3.1.6. *Internal agents*	72
3.1.7. *Genes as an "illusion"*	74

3.2. Second Transformation of the Person:
    Clients as Risk Carriers — 77
    3.2.1. A grave misunderstanding: Risk as diagnosis — 78
    3.2.2. The client as a statistical construct — 81
    3.2.3. The pathogenic effects of physician-attested risks — 86
    3.2.4. Life in irrealis mood — 89
    3.2.5. The genetic risk — 95
    3.2.6. The genetic self — 102
3.3. The Compulsion to Risk Management: The Decision — 105
    3.3.1. The imperative of the autonomous decision — 107
    3.3.2. The option requiring a decision: The test — 109
    3.3.3. Self-determined helplessness — 115
    3.3.4. Decision making:
        The paradox of personal risk assessment — 123
3.4. The Decision Trap — 134

4. Conclusion: Disempowering Autonomy — 140
    4.1. The Tyranny of Choice — 140
    4.2. Autonomous Decision Making as Social Technology — 143
    4.3. Conclusion: Now What? — 145

Transcription Conventions — 147

Bibliography — 148

Index — 168

Acknowledgment

I am grateful to many friends who have helped me to get the English book published: Mario Sagastume has patiently worked with me on the English book proposal and has helped me to get my ideas across in a language still foreign to me. Sajay Samuel and David Cayley have kindly revised my English and improved many paragraphs. Nancy Joyce has made the fluent and sound translation from German. And I thank Barbara Katz Rothman for her wise and beautiful preface.

Some material in this book had been adapted with permission from Samerski, S. (2014) Genetic Literacy: Scientific Input as a Precondition for Personal Judgement?, in Dakers, J.R. (ed.) *New Frontiers in Technological Literacy: Breaking with the Past*, New York: Palgrave Macmillan.

Barbara Katz Rothman

Preface to the English Edition

Choice, I have often pointed out, is not an alternative to social control, but a mechanism for achieving it.

This point is made, skilfully, dramatically, repeatedly, in this book. Decisions, also thought of as choices, are traps.

People enter these traps willingly. More than that—they eagerly seek entry. They are told decisions and choices are theirs to make, and all that is needed is rich and full information. With all the information, intelligent choices can be made, thoughtful decisions reached. Getting the information becomes a goal of its own, the path to proper decision making.

How can that be a trap? How can those choices be mechanisms for social control?

First, let's start with the simplest examples of choice used to control. Let's say you have a little child, and it is very cold and wet outside, and you want that child to put on a hat. "No!" baulks the child, "I don't want to!" Force is an option, but it won't be pleasant and it won't be quick. There may well be a temper tantrum, torn or thrown hats involved, a bad way to start the day. Information might work on its own, depending perhaps on the age and certainly on the mood of the child— explain how you might catch cold if you get cold and wet. It's been known to work, but... not so much actually in the real world. Pleading is rarely successful—"please, please be good and wear the hat" doesn't often get the hat on. Bribery sometimes work, "incentivizing" choices—wear a hat and you can have a treat later. But choice—choice may well work best of all. Would you rather wear the blue hat or the red hat? To make this technique work even better, offer the choice before you even raise the issue of hat-wearing, before the option of hat-refusal even enters into the child's mind. As the child pulls on a coat, or finishes the last of breakfast, ask what colour hats are there in the closet? Few children won't be caught in that trap: one hat is grabbed, the others discarded, personal autonomy saved, the day saved, and the hatted-child is on the way out the door.

Is it fair to equate a manipulated child to a modern, autonomous patient? Obviously not, and we are surely well beyond the era of the paternalistic physician. And yet, there are just so many ways anyone can get anyone else to do anything: force, pleading, incentivizing, and educating. In medicine, and most especially in the area of genetic medicine, and perhaps *most* especially in German genetic medicine, there will be absolutely no discussion of force or of incentive. The history of eugenics is just too close and too painful to even think about using any formal authority to require or to entice people into using any genetic technologies at all. Pleading works well when personal relationships are involved: one can well imagine a mother pleading with her daughter to use certain prenatal testing or a wife with her husband; within the confines of familial relationships, there is little doubt that pleading—personal urging, "because I care about you and you about me"—has a role. But pleading will not, cannot, work in the clinical setting. Which leaves us with education and its proffered choices. Clinicians will give information, all the information they can muster, describing every possible hat in every available closet, with great clarity and deliberation, leaving the choices entirely to the clients.

The big questions offered by the new genetics make a lot of people uncomfortable: you are going to have a baby—do you want to make some choices about the kind of baby you will have? You are going to die of something eventually—do you want to know more in your youth about what that is likely to be in your old age? These are things to which many people would say no, would turn away, these are the paths they do not want to walk down. That is not what is usually asked: instead we are asked which tests we will choose to have prenatally, which screening and predictive tests we choose for ourselves and our not-yet-born, maybe not-yet-conceived, children. We are taught more and more about the kinds of tests available, the differences between them, so that we can make intelligent and thoughtful choices, autonomous choices.

When people hesitate, when they push back against the choices, more and more, ever non-directive education is on offer.

And again, go to the simple choices in life and consider when and how we use education to structure choices. If you want someone to pick you up at a train station, you don't say "you do realize of course that you could have a fatal accident on the way? Here let me look up the numbers—these are the odds that you could lose control of the car and smash it; these are the odds that a little kid could dart out into traffic and you'd kill it; these are all quite small numbers, but... are you ok with taking that risk? Pick me up at 7?" Without making them think through all of what could indeed happen, without the actual risk numbers available to be explored, can you say they are making an "informed choice"? It sounds crazy, laughable, or perhaps sick to talk that way. But that is how "informed consent" counselling most often works. If the person wants what you want them to want, you give them whatever basic information might be required, and that's that. If they want something else, you keep feeding more information to be very sure they are making an "informed" choice. We educate people who are acting not in accordance with what we, the educators, think an informed and intelligent choice should be. We do not continue to educate people who are making what we understand to be good, thoughtful choices.

At every level, from large scale public education programmes through to individual counselling sessions, education is used to shape the kinds of questions that will be asked, to direct our attention to some issues and away from others, to some values and away from yet others. People who know that a lot of people in their family have had this or that disease have some sense that the disease could strike them too, is part of the fabric of their lives. But start educating them about the genetic basis for the disease, the tests that could tell them if that disease-causing gene is in their own bodies, and the experience of risk becomes different. The risk moves from "out there", in the world, or "on paper" in a family tree as one person says in this book, to an internalized part of one's own body, begins to move from "if" to "when". Dealing with risks that are "out there", that happen, that could happen to oneself is one thing; dealing with flaws in one's own body or one's baby's body, the genetic misprints the geneticist sees, is a very different project.

With education, with counselling, we move people from the one project — the inherent riskiness of life and the inevitability of death — to another: managing medical risks with currently available technologies.

It's a trap.

Preface to the German Edition

"Autonomy", "responsibility", "choice" — these are the big catchphrases that dominate policy making today. In every situation citizens are called upon to make their own choices, regardless of whether they have just become unemployed, pregnant, or had a genetic counsellor predict a scary disease in their future. A host of experts have made it their business to guide us in the decision making process. The business of professional counselling, which has turned deciding into a service product, is booming. A broad array of educational and counselling services exist for the sole purpose of guiding citizens on the path toward freedom and autonomy. Their assumption is that people only need the right "inputs" to handle an intangible and technologized world.

My scepticism over this professional encroachment on deliberation and decision has been growing for a long time. More than ten years ago I agonized for the first time over counselling interviews in my dissertation: counselling interviews in which genetic counsellors educate pregnant women about the possible chromosomal aberrations and risks of deformity their embryo carries and the relevant testing options, and then strongly implore them to make a "self-determined decision". How did it come about, I asked myself, that genetic counsellors whose aim for decades was to enhance the human race suddenly took it upon themselves to empower pregnant women? And what do pregnant women learn from this training in self-determination and decision making? Even now, I see genetic counselling, conceived specifically to help pregnant women decide whether or not to undergo genetic testing, as a paradigmatic example of the many different forms of decision making instruction citizens encounter today. I assume that elements and variants of the "decision trap", of which genetic counselling is a particularly blatant example, can also be found in other areas—fertility counselling, career counselling, educational counselling, and end-of-life counselling.

My reflections on the growing obligation to make informed decisions are fuelled by numerous encounters and conversations with Barbara Duden around her guest table. To the collegial circle that gathered there I owe an inner foothold and surprising insights—both prerequisites for pursuing my own line of thinking. Barbara Duden helped me adopt the critical distance to modern certainties that first enables fruitful reflection on the present. I learnt from her that people cannot "keep their wits about them" when their self-perception is being reshaped by scientific abstractions and bureaucratic categories. To Johannes Beck I owe my understanding of professional counselling as an educational event, as an example of the sweeping educationalization of every area of life. And Sajay Samuel helped me realize that the facilitation of risk-oriented decision making is imposing on clients a managerial rationality that originated in the so-called decision sciences in the middle of the twentieth century. In addition, a number of friends have contributed to the fruitful discussion about the consequences of genetics and prenatal testing, the citizen as a decision maker, the educationally needy person, and the loss of common sense. To them I owe many inspirational insights and realizations, and, not least of all, the impetus for putting my thoughts on paper in the form of a book.

This book is based on a research project. For two years Barbara Duden, Ruth Stützle, Ulrike Müller, and I deliberated on what happens when the "gene" is released from the laboratory into everyday life. What does the word "gene" mean in colloquial language? What does it say, suggest, and postulate when it emerges in a conversation between physician and patient, mother and daughter, or in a chat over the garden fence? Our research project, headed by Barbara Duden in the Institute of Sociology at the Leibniz Universität Hannover, was titled "The Everyday Gene: The Semantic and Praxeological Contours of the 'Gene' in Everyday Speech" and received funding from the Federal Ministry for Education and Research (BMBF). The project provided me with the opportunity to observe additional genetic counselling sessions and to explore further the relationship between the scientification of everyday life and the educationalization of decision making.

I owe a big thanks to the genetic counsellors who allowed me to observe their work as well as to their clients who agreed to my presence. Another thank you goes to BMBF for their generous funding. A real godsend were those who read my manuscript, which benefited considerably from their clever remarks and comments: Frank Butzer, Barbara Duden, Friederike Gräff, Marianne Gronemeyer, Ludolf Kuchenbuch, Thomas Lösche, Ansgar Lüttel, Antje Menk, Uwe Pörksen, and Gudrun Tolle. Another godsend, and no less indispensable, were the friends who helped us manage our everyday lives while juggling the book manuscript and little children. A special thanks in this regard goes to Michael Lienesch, Johanna Germer, Antje Menk, Ina Sapiatz, Dorothee Torbecke, the grandparents, and the team at the nursery school Picobello e.V.

I would like to dedicate this book to Ivan Ilich and my daughters Hannah and Alena. The former gave me the hope and thus the courage to confront modern myths. To my children I owe my sense of what is truly important. I would also like to give them hope on their paths.

Chapter One

Introduction

Gene as the Basis for Decision Making?

The non-squashy tomato from the genetic engineering laboratory is called a "GM tomato" in popular parlance and the manipulated food of the genetic engineering industry "GM food". Surveys show that many people assume that these "genetically modified foods" contain genes while tomatoes from their own gardens do not.[1] Geneticists and enlightened journalists appear to be indignant over this alleged ignorance of the population. The majority do not know that they are always chewing genes, lamented a commentator in the *Frankfurter Allgemeinen Zeitung* (Müller-Jung, 2006). And Hanoverian plant geneticist Hans-Jörg Jacobsen questions whether such ignorant citizens can even participate in a democracy. One asks inevitably, he claims, how our democratic society can make decisions on this kind of meagre and uninformed basis. A proper public discourse is called into question and the floodgates are open for one-sided ideologically-based agreements (Jacobsen, 2001).

---

[1] The survey Eurobarometer asked this question for several years, see Eurobarometer (2006, p. 57). Brian Wynne criticizes this kind of survey for degrading common knowledge as ignorant, see Wynne (1995).

This lament over an unenlightened population and the necessity of disseminating "proper" information is widespread. In the age of genetics, science, politics, and industry want citizens who can participate in biopolitical debates and make "informed decisions". They wish to remedy this bemoaned ignorance of the population through information and education. Today's citizens are not only to learn reading, writing, and counting, but also to cram the genetic ABC. "Genetic literacy" is the buzzword that evokes this new educational goal. Genetics taught in schools, in public laboratories, and formal discussions should empower citizens to see the world through the lens of a geneticist. They should know that tomatoes, like people, carry genes, and should base their decisions on these facts.

This attempt to educate citizens so they can make "informed choices" in the sphere of genetics is a theme of this book. I want to challenge the assumption that genetic education empowers anyone to make an independent, autonomous decision. Rather, it seems to me that genetic education subverts a freedom: the freedom to know and decide for yourself without guidance. The once emancipatory call to not be patronized but to be guided by your own intellect has been subverted into the duty to make an "informed choice". Today, in every situation in life people are expected to call on professional services to turn themselves into informed decision makers: health insurance companies exhort their customers to exercise their personal responsibility and offer patient education and decision making aids; the pharmaceutical concern Roche wants citizens to become responsible health care managers (Höhler, 2009); discourse projects provide training for participation in biopolitics; and physicians determine individual health perceptions and treatment preferences for their patients. When Immanuel Kant (1724–1804) in 1794 articulated "Sapere aude!" (dare to know) as the battle cry of the Enlightenment, he could not have foreseen the emergence of an educational and counselling industry. At that time there was no "knowledge society" in which industry-produced information[2] was sold as tangible knowledge. He could not have suspected that one day a whole army of experts would thrust their services upon citizens in order to enable them to make so-called informed choices. Nowadays, terms such as "enlightenment" and "self-determination", which Kant invoked in his call for emancipation from conventional authorities, have degenerated to the catchphrases of a new social technology.[3]

A number of governmentality studies have shown that the prevalent forms of exercising power are no longer coercion and repression, but rather the steering of self-determination or guidance for self-guidance (see, among others, Bröckling, Krasmann and Lemke, 2010; Rose, 1999). Based on Michel Foucault's work on modern techniques of governing, they analyse the mobilization toward "information", "responsibility", and "self-determination" not as emancipatory progress but as a new form of governance. These studies have helped me in examining as a form of social control the call to make "informed choices" on the basis of genes and risks. I could ask how genetic education and genetic counselling function as a social technology. What is being asked of citizens when "responsibility" requires having experts update your own powers of judgment? Doesn't the alleged "proper" information

---

[2] Philologist Uwe Pörksen ranks the colloquial word "information" among a new word class, namely the plastic words (Pörksen, 1995). Since "information" has advanced to a technical term in the information sciences and then migrated back to colloquial language, it lost its colloquial complexity and flexibility as well as its scientific denotation. It has mutated into a connotative stereotype, a word that sounds meaningful but says nothing. Along the same lines, Lily Kay has called the word "information" in the life sciences, where it has migrated from the information sciences, a catachresis: "Emptied of its technical content, it actually became a metaphor of a metaphor, a signification without a referent. This, however, did not diminish its scientific and cultural potency" (Kay, 2000, p. 127).

[3] Social technology means the planned and scientifically based control of the social and is partly used synonymously with the terms social control or social engineering.

already provide the bottom line, namely the framework for deliberations while generating the decision options? In what form of thinking are citizens being initiated when they are no longer supposed to be moved by experiential, tangible, and colloquially comprehensible realities, but rather by scientific constructs such as "gene" and "risk"?

What I want to study could be called the symbolic power of genetic education. I ask what the instructed learns beyond the imparted specialized knowledge—about themselves, their being-as-it-is, their judgment, and their duties as citizens of a technological society. I can best analyse this symbolic power at the point where expertise directly meets everyday life: that is, where the direct convergence of genetics and the biographical, service products and the desires of daily life, scientific jargon and colloquial speech,[4] technological rationality and personal meaning, happens. This place is the genetic counselling session. Genetic counselling is a professional service that aims to educate laypeople to become decision makers in matters involving "genes" and "risks". In a one to two hour session a genetics counsellor teaches his clients about DNA structure, heredity laws, mutations, disease incidence, statistical risk calculations, and genetic testing options. Such a counselling session is a classic example of the attempt to train citizens to become responsible decision makers in matters pertaining to genetics. The explicit goal of the education is "informed choice" — typically a decision over whether to have oneself or an unborn child be genetically tested or not. Meanwhile I have observed, edited, and transcribed nearly forty of these conversations; in

---

[4] Colloquial language, even in its highest abstractions, is based on experience and sense perception: "The categories and ideas of human reason have their ultimate source in human sense experience, and all terms describing our mental abilities as well as a good deal of our conceptual language derive from the world of the senses and are used metaphorically" (Arendt, 1963/2007, pp. 47–8). The criteria for differentiating between scientific language and everyday language I take from linguist Uwe Pörksen (1986; 1995).

five of them a woman or a couple are being prepared to make the decision as to whether they want to undergo genetic testing to determine their risk of cancer. In the others women are being educated about pregnancy risks as well as the option of having their unborn child genetically tested.

Distancing as a Research Approach

How can the symbolic power of a genetic educational event be studied? How can I identify the hidden thought constraints, suggestions, and subliminal demands in a genetic counselling session? The present can best be studied when I distance myself from it. If my analysis is based on the same assumptions as the object of my study, then it is in danger of merely reinforcing existing truths. If I marvel at "genetic knowledge", believe in statistical prediction of the future, and consider "informed choice" to be the epitome of autonomy, I will not be able to grasp the contemporary significance of genetic education. If I confound a correlation with the identification of a cause, if I confuse a statistical risk with the tangible danger, or if I view an unborn child as an intra-uterine gene carrier, then I already share the worldview of genetics: I concede to genetics precisely the interpretive authority over reality that I want to make the object of my analysis.

My dual-track education as a natural scientist and social scientist has helped me adopt a critical distance to the modern faith in genes. It has saved me from two traps that lurk when studying the "release of genetic concepts" into everyday life: I do not consider "genes" and "risks" to be objective facts or even natural objects, nor do I consider them to be terms whose meanings—as with colloquial words—can be defined by the speaker or negotiated in the conversation. While studying biology I grappled with the peculiarity of the formation of biological concepts, and my work in cytogenetic and molecular genetic laboratories has rid me of illusions about the production and dissemination of what is called genetic knowledge. During this time I crisscrossed between the disciplines of biology, philosophy, and social sciences; and the travel between laboratory and the reading of Ludwik Fleck, Lorraine Daston, and Barbara Duden has proven to be extremely fruitful. I realized how hypothetical and context-dependent scientific facts are, how precise and discipline-specific their validity is, and how much they change when they emigrate to everyday life. Back then I was already participating in genetic counselling sessions out of curiosity and prepared the first session transcripts. I was interested in the question of how a laboratory construct becomes a portentous verdict in an educational conversation. A modified DNA sequence, that is, a molecular variation, becomes a fatal "gene for cystic fibrosis" and calls into question the birth of a child; a statistical frequency mutates into a risk that gives a pregnant woman sleepless nights. From my multidisciplinary studies I took two fundamental insights: first, I realized how much scientific research findings are determined by biases, experiment structure, third-party funding stipulations, and endeavours to build careers; and second, I realized the chasm between the laboratory and the everyday world as well as the artificial worlds and pseudo-realities that can emerge when we attempt to build a bridge between them.

It's not only my dual-track education, however, that has provided me with some distance to modern certainties. The more I struggled to understand the peculiarity of genetic predictions of the future, professional educational rituals, and the duty to make an "informed choice", the more I saw a critical need for a historical awareness of our time. I started to question genetic educational and counselling events in an entirely new way when I learned that no professional counselling industry existed before the twentieth century and that its expansion occurred only in relation to the growing power of experts in the welfare state. Likewise, I can only comprehend how strange the empowerment of physicians and geneticists over the care of an unborn child is when I know that a few generations ago a pregnant woman was "expecting" a child to come and was by no means the "uterine environment" for a "fetal development". Furthermore, if I am not a historian and hence have no "mooring" in history—the methodological distancing, that is, the attempt to "be estranged from the present in looking at it from the perspective of the past" (Duden, 2002, p. 7), is an indispensable foothold for my thinking.

Education about "genes" and "genetic risks" seems natural as long as I believe in the universality and significance of what is being conveyed as "knowledge". To analyse the symbolic power of genetic education, I must be able to free myself from my belief in genes. Aiding me here is the growing challenge to the concept of the gene in science. In the following chapter I will briefly address the scientific and social significance of the gene to make it clear that the "gene" is not a natural object, not even a scientific fact, but rather is a scientifically supported myth. Subsequently I want to describe various campaigns and endeavours to provide genetic education for the public and to record my initial observations of this pedagogical programme. In the third chapter I will finally let my readers participate in educational conversations in which citizens are being empowered with a knowledge of genetics. I will take readers to genetic counselling agencies where genetic counsellors educate women and couples about genes, risks, and their options. Taking these sessions as an example I will examine step by step the symbolic power of the genetic educational and counselling events which guide citizens to make "informed choices".

Chapter Two
Genetic Education

2.1. The Gene

An era typically receives a name when it comes to an end. Only when self-evident truths are challenged does the singularity of a specific period become apparent. Evelyn Fox Keller named the twentieth century "The Century of the Gene" and wrote an informative book about the rise and fall of genetic thinking in biology. Invoking a growing number of biologists and geneticists she concludes that the gene as a heuristic notion is obsolete. The thesis of a definable, controllable, and causative gene is scientifically antiquated (Keller, 2000). The geneticist Wilhelm Johannsen's criticism of his colleagues' belief in genes is as penetrating today as it was nearly one hundred years ago. In 1909, he launched the term "gene" and soon found it necessary to dampen the rampant gene euphoria escalating around him. The idea that there are discrete trait-defining hereditary units must be abandoned as not only naïve but also completely misleading, warned Johannsen in 1913 (Johannsen, 1913, p. 144).

Johannsen's objection was ignored at the beginning of the twentieth century. But today, nearly one hundred years later, his protest has become a burning issue once again. Increasingly more geneticists are arguing that they can no longer justify talking about a gene as a discrete segment on a chromosome, as a sufficient cause of disease, as a building block of an organism, as a carrier of "information", or even as a hereditary unit or functional unit.¹ Particularly the findings of the Human Genome Project have clearly shown even die-hard gene determinists the confusing complexity of development and heredity. The notion of the gene as a causal agent has been definitely shaken (Keller, 2000). New research shows that the gene for alcoholism, low intelligence, dementia, or a thick belly draws on a series of fictions. To understand biological interactions, concludes Kelly, the concept of the gene is not only just outdated, it has even become a hindrance.

Scientific concepts and theories come and go—this insight no longer astounds anyone. As early as the nineteenth century the belief in the progress of knowledge and the finality of scientific findings showed its first cracks. At the end of the nineteenth century the French mathematician and physicist Henri Poincaré suffered from vertigo, as did many of his colleagues, when he realized how short-lived scientific theories are, how many new observations are made every day, and how contradictory and irreconcilable these often are (Daston, 2001, p. 213). A few decades later, in the 1920s and 1930s, the physician and philosopher of science Ludwik Fleck, in a pioneering work that still resonates today, analysed the "Genesis and Development of a Scientific Fact" (Fleck, 1981) as a social event, thus paving the way for a socio-cultural historical understanding of scientific knowledge. Fleck demonstrated that a scientific fact is a socially constructed "thought constraint" (Fleck, 1981)—an attempt at understanding that gradually condenses to a truth and then can be thought of only this way and no other way. Drapetomania— a slave's pathological compulsion to flee—Freudian libido, female hysteria, as well as spirochete as the cause of syphilis and the normal distribution in statistics, presuppose a specific socio-culturally bound perceptual disposition to be conceivable and plausible. This unconscious bias in thinking, this directed perception with the corresponding cognitive and objective

---

[1] See Beurton, Rheinberger and Falk (2000), Burian (1986), Falk (1984), Rheinberger and Müller-Wille (2009), Strohman (1997), among others.

processing of the perceived, is what Fleck calls thought style. If the thought style changes, scientific objects previously regarded as rock-solid facts either acquire an entirely new meaning or simply vanish.[2]

The "gene" is nothing more than such a socially and culturally constructed "thought constraint" that has held sway over biologists, physicians, bioethicists, and research policy makers for several generations. As early as the nineteenth century, before Mendel's rediscovery seemed to deliver the final proof for the existence of hereditary units, biological thinking was attuned to the idea of "genes". Numerous scholars, from Charles Darwin (1809-1882) to Francis Galton (1822-1911), August Weismann (1834-1914), and Hugo de Vries (1848-1935), had already postulated hereditary particles, and Mendel appeared to be providing proof of what was regarded as self-evident and obvious.

Nowadays, on the other hand, in the age of "dynamic networks", "interaction", and "irreducible complexity", determining genes are irrevocably obsolete. The thought style of our age, characterized by cybernetic or systems theoretical axioms and concepts, has finally settled into genetics and there made the "gene" an emergent construct of the cellular and organismic system. Today everything that has governed scientific thinking about life for an entire century is obsolete.[3] We find ourselves at a "watershed" in biology (Beurton, Rheinberger and Falk, 2000,

---

[2] Thus even scientific objects have a "biography", see Daston (2000).

[3] Everything I learned for my genetic studies at the University of Tübingen through 1989 is now obsolete: the "central dogma of molecular biology", that all information comes from DNA, is just as obsolete as the so-called "one gene-one enzyme hypothesis" (genetics now knows DNA sequences that can code for thousands of proteins). What we were taught to regard as "junk" DNA, that is, as evolutionary garbage, is today considered to be the site of highly complex gene regulation. And RNA is no longer merely a passive messenger, but has been revealed to be a governing molecule that actively transforms genes and gene activity or gene products.

p. ix), of which many biologists, not to mention the general public, have not taken note.[4]

And at this watershed the gene, like other scientific facts, whether it be the phlogiston[5] or Bohr's atom model, has a history with a beginning and an end. For a while it was a plausible and useful concept that condensed to a scientific fact until it has had to yield to new hypotheses and theories.

Unlike the phlogiston or Bohr's atom model, however, the gene has had an adventurous career within as well as outside science. It has governed not only research and theories but has also determined policies, turned medicine inside out, and underlies fundamental cultural beliefs. Genes, as geneticists have led us to believe, are indispensable to life. They determine the development of a person, their personal being-as-it-is, their health, and their illnesses. According to the gene creed, genes

---

[4] The anthology by Beurton, Rheinberger and Falk (2000) summarizes the most important scientific historical and theoretical reflections on the "gene". In the preface the editors confirm the occurrence of a radical upheaval in biological thinking: "Somewhat detached from the gene as a public icon, but also unknown to many biologists, these new findings have caused a watershed during the last few decades. The more molecular biologists learn about genes, the less sure they seem to become of what a gene really is" (Beurton, Falk and Rheinberger, 2000, pp. ix–x). Genes do not cause or determine, they do not contain instructions, they are not independent of their cellular or organismic environment, they are not isolable and not stable—indeed, they cannot really be defined: "Rather than ultimate factors, genes begin to look like hardly definable temporary products of a cell's physiology. Often they have become amorphous entities of unclear existence ready to vanish into the genomic or developmental background at any time" (Beurton, Falk and Rheinberger, 2000, p. x). The entry for "gene" in the Stanford Encyclopedia of Philosophy is up-to-date and insightful, see Rheinberger and Müller-Wille (2004).

[5] The phlogiston is a hypothetical substance, which was believed in the eighteenth century to be the cause of combustion. Between 1700 and 1789 it was regarded as a scientific fact, and although it was finally proven to be a phantasm, it was very successful as a research construct for almost a century and served to advance the cause of modern chemistry. On the phlogiston as a "scientific fairy tale", see Pörksen (2003).

cause lumps in breasts, nervous nail-biting, and a fat belly (Duden and Samerski, 2007); they bond the mother to her child and are responsible for male infidelity; they predict the diminished IQ of an unborn child and the coming infirmity of a 20-year-old. And as much as the gene seems to explain, just as promising is its research and manipulation: as the building blocks of genetic technologies it augurs bacteria that eat toxic industrial waste and well-fed African children playing in green deserts; and Hoffmann-LaRoche laboratories promise hope for individual health cocktails and new panacea against cancer, infirmity, aging, and death. The biochemist Erwin Chargaff sums up the modern faith in genes: "Today genes are everything. A fundamentalist belief prevails" (Chargaff, 2001, p. 249).[6]

This faith in genes, however, is not simply an inevitable side effect of genetic research, but is its existential basis. The defining, diagnosing, and patenting of genes is big business. The pharmaceutical and agro-industries, scientific research, and the growing service market for prenatal diagnostics, genetic testing, and bioethics live on the belief in genes.[7] Geneticists admit that previous concepts of genes were "naïve" (Stefan Klein and Venter, 2009), but from this insight conclude nothing

---

[6] Cultural scientist José van Dijck describes this "biophoria" as the result of a multi-million dollar perception management: "The 'geneticization' of society seems to be the flip side of the 'medicalization' of genetics. Despite its abstract theoretical goal, human genome mapping managed to generate overwhelming enthusiasm among the general public. A true 'biophoria' in popular representations of the genome project resulted from a strong injection of powerful images and imaginations in the public domain..." (van Dijck, 1998, p. 120).

[7] "The presumption that genes operate independently has been institutionalized since 1976, when the first biotech company was founded. In fact, it is the economic and regulatory foundation on which the entire biotechnology industry is built" (Caruso, 2007). A molecular biologist from New Zealand speaks of the "industrial gene": "The industrial gene is one that can be defined, owned, tracked, proven acceptably safe, proven to have uniform effect, sold and recalled" (Heinemann, cited after Caruso, 2007).

more than the need for more genetic research. They are in search of a new form of genes: probabilistic genes or "susceptibility genes".[8] Based on a deluge of data produced in genetic laboratories, they construct statistical correlations[9] between genotypical and phenotypical[10] traits which are then interpreted as susceptibilities or dispositions. Hypotheses on pathogenesis and aetiology are not necessary. As bioinformational[11] constructs these "genes" stand for purely statistical relations. On

---

[8] In contrast to the colloquial meaning of susceptibility, a genetic susceptibility or disposition for cancer or dementia does not indicate a bodily constitution. Genetic susceptibility is a purely statistical construct. The susceptibility gene for Alzheimer's disease, for example, is an acronym for a correlation between the frequency of DNA variations and the frequency of the disease. The so-called "gene for" Alzheimer's is neither a cause nor a precondition for getting it. Most Alzheimer's patients test negative, and most gene carriers do not get Alzheimer's (numbers diverge widely in different studies, see, among others, Lock et al., 2006, pp. 283–5).

[9] "Correlation" is a technical statistical term denoting the stochastic dependency of two statistical events. If there is a correlation, the two events or variables have a statistical connection that can be mathematically modelled. Statistically seen, they do not appear independently. Thus, a correlation disproves the null hypothesis that assumes stochastic independence.

[10] Genotype and phenotype are, in contrast to their popular scientific understanding, classifications. "The 'phenotype' of an organism is the class of which it is a member based upon the observable physical qualities of the organism, including its morphology, physiology, and behaviour at all levels of description. The 'genotype' of an organism is the class of which is member based upon the postulated state of its internal hereditary factors, the genes" (Lewontin, 1992, p. 137). Thus, genotype and phenotype can be correlated, but it is not possible to extrapolate from genotype to phenotype in a single case: "As is true for living systems in general, relations between genotype and phenotype are contingent, varying from case to case" (Lewontin, 2004).

[11] Bioinformatics, also called computational life science, is a discipline that processes biological data with the help of computers. Today, bioinformatics is a foundation of the life sciences. Genome research produces large amounts of data about DNA, proteins, gene expression, RNA, and so forth that are stored in big databases and statistically analysed by software programs. These programs model molecular structures, compare sequences, construct gene regulation models, or associate genetic markers with phenotypic features.

this basis geneticists continue to announce the discovery of "genes for", whether it be for aging, homosexuality, speaking, smoking, or the "God gene".[12] They are promoting a new genetic pandeterminism in which the gene is no longer regarded as the sole cause but as an alleged co-agent that plays a role everywhere—whether early-onset dementia, school failure, the flu, or suicide.[13]

Even though the concept of the gene is obsolete in research, talk about genes has certainly not died down. The gene as a shorthand term in technical discussions is too practical, as a didactic aid too appealing, as a moneymaker too profitable, and as an instrument of propaganda too effective to discard. That genes are useful not only in biology but also in politics and society is not a new phenomenon. Even before the existence of hereditary units was considered to be proven, the predecessor of the gene was already fuelling feasibility fantasies and promises of salvation. Heredity scientists such as Francis Galton or Alfred Plötz (1860–1940) not only endeavoured to clarify questions in science at the end of the nineteenth century, they were also striving for a new order in society. They wanted to use their genetic expertise to contribute toward the creation of a scientifically based social order. The "Mendelian units", which Johannsen named "genes" in 1909, pushed the vision of a rational reproduction policy within reach, which no longer left starting a family and childbearing to the individual. Heredity

---

[12] The Gene-ethical Network (GeN) offers a small collection of various "genes for", see http://www.gen-ethisches-netzwerk.de/gen-fuer [31.8.2014]. The same geneticist who announced the "God gene" made quite a splash in the media in 1993 with the claim that he had discovered the "Gay Gene", see Hamer (2006).

[13] Regardless of whether flu or suicide—genes today play a role in everything: "In the United States, 9 out of 10 leading causes of death have known genetic components: heart disease, cancer, stroke, chronic obstructive pulmonary disease, pneumonia/influenza (through the immune system), diabetes, suicide (through tendency to depression), kidney disease and chronic liver disease. Together, they account for 76 percent of deaths in the USA. At the same time, genetics offers hope for effective treatment or cure" (Wertz and Fletcher, 2004, p. 284).

researchers presumed that those who did not meet the demands of industrial society had abnormal or damaged genes and dismissed them as biological waste. Whether vagabonds, slackers, the infirm, crippled, or other eccentrics — those who disturbed the image of a modern, rational society were labelled genetically inferior and were either to be adapted with the help of medical and educational measures or eliminated through confinement, sterilization, or euthanasia.[14]

The days in which the gene served to stigmatize and eliminate unpopular ethnic groups are gone. Today the gene is no longer a eugenic instrument of authoritarian population policies but rather an instrument to mobilize citizens. It has been praised as the basis of an active, self-empowered way of life.[15] According to science, politics, and industry, genetic literacy is a *sine qua non* for rational self-governance and democratic citizenship.

---

[14] Michel Foucault has argued that modern societies are shaped by technologies of normalization: the human sciences, whether medicine, psychology, or sociology, construct a (statistically defined) norm by which people are measured. Deviations from this norm are considered to be pathological and thus in need of treatment. Genocide, the murder of ostracized and problematized ethnic groups, can be an extreme measure of such a "treatment", whose goal is to create a scientifically based social order, see Foucault (1990a). In his analysis of the Holocaust as an outgrowth of rational social planning, Zygmunt Bauman (1989) also points out that modern societies which desire to create an artificial social order are inherently racist. Those who do not fit into this envisioned order become a problem that must be solved by means of social planning, medicine, and in extreme cases, extermination: "one can and should remake society, force it to conform to an overall, scientifically conceived plan... This is a gardener's vision, projected upon a world-size screen... Some gardeners hate the weeds that spoil their design — that ugliness in the midst of beauty, litter in the midst of serene order. Some others are quite unemotional about them: just a problem to be solved, an extra job to be done. Not that it makes a difference to the weeds; both gardeners exterminate them" (Bauman, 1989, p. 91).

[15] Genetics has become the basis of active self-governance. Genetic terms and concepts determine everyday deliberations and decisions and create new responsibilities; see, among others, Rose (2007), Novas and Rose (2000), Bunton and Petersen (2005), Kerr (2004), Petersen and Bunton (2002).

Only citizens well-versed in genetics can be responsible citizens capable of taking destiny into their own hands.

Various genetic educators, from geneticists, industrial representatives, to science journalists, communication experts, and even social scientists, ethicists, and advertising agencies, strive to inform citizens about the inner life of their cell nuclei and instruct them in weighing opportunities and risks. On the one hand, according to the rationale behind this mobilization, in a few years genetically modified food and genetic testing will become commonplace in everyday life. And in a geneticized[16] society those who do not want to be left behind must become well informed about DNA, genetic technology, risks, and ethical dilemmas. On the other hand, as many geneticists and other believers in genetics promise, genetic education leads to self-knowledge and self-determination. Thus, in the same way that all tomatoes are genetic food, they say all people are carriers of genes. Therefore "genetic knowledge" is regarded as a *conditio sine qua non* for self-determination and autonomy. "Respect for autonomy actually leads to... the obligation to pursue genetic knowledge", postulates, for example, the ethicist Rosamund Rhodes (1998, p. 17). In their perspective, the terms "self-determination" and "autonomy" stand exclusively for the obligation to make genetically "informed decisions".[17]

---

[16] Abby Lippman coined the term "geneticization" (Lippman, 1991), with which she identifies the increasing reference to "gene" in quite different social contexts, whether medicine, crime, or something else: "Geneticization refers to the ongoing process by which priority is given to searching for variations in DNA sequences that differentiate people from each other and to attributing some hereditary basis to most disorders, behaviours, and psychological variations" (Lippman, 1994, p. 13).

[17] "Autonomy is viewed not as a natural given but as an achievement that demands that the subject be fully 'informed' about their susceptibility to risk and about available options, the assumption being that more genetic information will create more choices", commented Petersen (2002, p. 139) on this new autonomy. A blatant example of the propagation of self-determination and "empowerment" through genetic education is the popular scientific brochure "Your Genes, Your Choices: Exploring Issues Raised by Genetic Research" (Baker, 1997), published by the

2.2. Educational Campaigns

2.2.1. Illiterate citizens? A Bremen congress

The Federal Agency for Civic Education (Bundeszentrale für politische Bildung/bpb) is a German institution that promotes citizen participation in politics. After the turn of the millennium, it saw the dawning of an era of genetic technology and considered it their duty to prepare the general public. Thus, it organized a national congress for civic education titled "Good Genes—Bad Genes?" that was held in Bremen in early September 2003.[18] For three days representatives from science, politics, and industry debated the "opportunities and risks of gene technology". The congress announcement asserts that the biosciences and biotechnologies are currently learning to understand, control, even improve fundamental life processes at a tremendous speed (bpb, 2003). In view of this invoked feasibility the sponsors of the civic educational congress hoped to enable citizens to be an active influence on the decision making process. To this end they invited to Bremen three dozen high-ranking experts from around the world: an Israeli molecular geneticist, with whom former German Minister of Justice Hans-Jochen Vogel discussed the ethical defensibility of stem-cell research; a physician from Nicosia, who reported on the church and state-implemented eugenics programme in Cyprus; an American historian from Philadelphia, who subsequently defended the Cypriot compulsory programme; and a human geneticist from Leuven in Belgium, who wanted parents to assume "the responsibility for their children's genetic make-up".

The congress was accompanied by a thematic cultural programme with films and readings. Prior to the start of the con-

---

influential scientific organization *American Association for the Advancement of Science* and financed by the US Department of Energy.

[18] The descriptions and verbatim quotations from presentations are based on records I prepared as a participant observer of the congress.

ference, the organizers indulged in a prank in Bremen's Ostertor district: a fictitious genetic shop called "Chromo Soma" opened its doors shortly before the congress started. Curious passersby were offered numerous genetic products and services: customers, for instance, could purchase "Gene Maxx®" to revitalize buried genetic potential and intensify their zest for life, or "Book a Baby®", which allowed customers to create their desired embryo and temporarily store it until a convenient time.[19]

The mission of the Federal Agency for Civic Education is to provide "citizenship education and information on political issues for all people in Germany. The work done by the bpb centres on promoting awareness for democracy and participation in politics" (bpb, 2012). The organizers also pursued this goal with the three-day congress on genetics. Their aim was to bring the topic of gene technology out to the general public for debate. A public discussion, declared the final plenum, would greatly contribute to the education and emancipation of the public, which would support citizens on the path to a reflected judgment (bpb, 2003).

This project, to "emancipate" citizens in the matter of genetics and to spur them toward participation, received the full support of science and industry: at the congress opening in Bremen's Marriott Hotel not only the organizers but also geneticists and pharmaceutical representatives complained that the general public was lagging behind in issues related to genetics. They would not take part in deciding where and how gene technology should be used. Particularly women and other affected citizens, it was said, simply refused to speak up. The three-day sessions with genetic and bioethical experts were to be a first remedy. For their part, the experts were clear about the cause of this lack of democratic participation: their diagnosis is that people are hopelessly backward where genetics is concerned. When it comes to DNA, heredity, and genetic testing,

---

[19] The bpb (2003) documented the event on the internet in German.

most people are ignorant and disempowered. They have opinions, but, for the organizers and the experts, the *vox populi* is too unqualified. Perceptions, complained one of the organizers, but not knowledge is what currently shapes the attitude of citizens toward gene technology. And the director of Bremen's Centre for Human Genetics attributed reservations about his field to misinformation: unrealistic hopes, he declared succinctly, lead to unrealistic fears. Therefore he prescribed counselling and education for his fellow citizens to enable them to deal rationally with genetics.

The experts saw their convictions confirmed by current reports coming from the genetics shop Chromo Soma: about one third of the customers believed in and purchased the fictitious products, reported the disgruntled shopkeeper. The level of knowledge among customers is alarmingly low, declared the shop manager indignantly, complaining about an enormous educational gap. In particular, he complained about the fact that his customers could not differentiate the fictional from the feasible and had exorbitant notions. Consequently, he also called for education and counselling: the fictional part, so it was said, should be dispelled in favour of a correct appraisal and assessment of genetics. (Ironically, what the shopkeepers considered fictional in 2003, namely the offer to "book a baby" for career women who want to delay childbearing, became reality only a few years later: so-called "social freezing" or "egg freezing", that is the freezing of women's "fresh" eggs at an early age for *in vitro* fertilization in later life. Thus, the boundary between the factual and the fictional is not that clear.)

Not just science and state educational institutions but also industry insisted on the need for an educated citizen able to participate in "democratic biopolitics". What they hope to gain from this was made very clear by the pharmaceutical representative from Roche toward the end of his presentation: society, not industry, bears the "responsibility" for gene technology, he explained. Society must decide what it wants to do with gene technology, he demanded. In his presentation it quickly became clear, however, that industry does not intend to simply subject itself to the will of the people. So far, he criticized, people have lacked the "knowledge" and the "assessment opportunities". Society, he made clear, must first be prepared for this new challenge. He literally said: "We must help society understand... We must explain to society how it should understand and how it must decide."

The congress did not achieve the goal set by the Federal Agency for Civic Education. On the contrary, for the most part a public debate failed to materialize. No discussion took place with the citizens attending the congress. Only a few members of the audience spoke after the presentations. The atmosphere in the hall was largely subdued, even paralysed. Apparently most listeners were rendered speechless. Although the presentations from experts were intended to stimulate discussion, they were not delivered in a manner that would be broadly comprehensible to laypeople, but instead were peppered with specialized professional jargon. The speakers talked about "zygotes" before and after "nuclear fusion", about "chromosome aberrations", "genes for" various unknown diseases, as well as "disease probabilities", "genetic dispositions", and "risk carriers". It was assumed that the starting point of a democratic discussion is not common sense but the scientific terms used in the discourse of experts. Discussed at the conference was not what Bremen citizens experience, fear, and desire, but scientific laboratory constructs and bioethical problems. Many presentations only touched on individuals and their experiences when the geneticists resorted to commiserating tales of woe to plead for research money and deregulation.

The Bremen congress is an impressive example of the attempt to educate citizens so they can attain greater self-determination in genetics-relevant issues. However, instead of dismantling barriers to democratic participation, the congress set up new ones. The bpb convened experts who explained that their professional knowledge is an essential precondition for a democratic discussion and that the entire population is consequently in need of counselling. Only those who are instructed by geneticists and bioethicists should have a voice about gene technology—in regard not to scientific, but to social issues. The proposed topic of the congress was not the scientific perspective on gene function and DNA structure; it was supposed to be about the effects of a new technology on society.

And precisely here is where the speakers denied their fellow citizens their autonomy. And in doing so they undermined any democratic discussion. Speakers without any genetic or bioethical schooling were declared incompetent. In fact, the congress agenda made no provision for them. Nor did the organizers give voice to the year-long protest by residents and the ecology-minded against the release of genetically manipulated corn or the "No Thank You!" campaign by women and the disabled movement against prenatal selection. Instead their keynote speakers were representatives from science and industry, who demanded that citizens' thinking and actions be adapted to the specifications of gene technology. They only wanted citizens to participate in "democratic biopolitics" after they learned to base their decision making on scientific input.

2.2.2. The genetic literacy campaign

In the twenty-first century, experts announce ever more educational goals. In order be fit for the technological society, people are not only to acquire literacy in reading, writing, and arithmetic, but also computer literacy, statistical literacy, media literacy, neuro literacy, health literacy, and, last but not least, genetic literacy. "When centralization and specialization grow beyond a certain point, they require highly programmed operators and clients. More of what each man must know is due to what another man has designed and has the power to force on him" (Illich, 1990, p. 58). The assumption is that those who do not want to be left behind by genetic knowledge production and genetic technology must cram the genetic ABC. Citizens of the twenty-first century, as numerous social scientists have suggested, are "genetic citizens". They see themselves as gene carriers, know the Mendelian rules and the structure of DNA, assemble with other gene carriers, think biopolitically, and make informed decisions (Heath, Rapp and Taussig, 2004). In this world of genetic citizenship, however, genetic literacy is key. It is the precondition for citizens and consumers to make informed decisions (Jennings, 2004). A number of institutions have thus taken on the mission of promoting the genetic literacy of the population. Self-help books, science centres, and science museums,[20] discourse projects,[21] citizen conferences,[22] life science learning laboratories in schools and on wheels,[23] physician–patient informed-consent discussions, and genetic counselling agencies all try to turn genetic illiterates into empowered citizens. As different as the educational programme and the public may be in each case, they share one goal in common: they strive to mobilize citizens by means of professional instruction to engage with "genes" and "risks" and become qualified for participation in informed debates and making informed decisions.

Such scientifically based educational projects are supported by policy makers and industry. The life sciences, and genome

---

[20] In Germany, science centres have sprung up like mushrooms, e.g. the Universum in Bremen, the Phänomenta in Bremerhaven, Flensburg, Lüdenscheid, and Peenemünde, the Spectrum Berlin, and many more. Traditional museums such as the Deutsche Hygiene-Museum in Dresden and the Deutsche Museum in Munich have started to offer hands-on science. Barbara Duden criticizes the visual infotainment of these exhibitions as pedagogical instruction in self-disembodiment, see Duden (2002a, pp. 187–99).

[21] The Federal Ministry of Education and Research in Germany supports numerous "discourse projects" or "discussion projects" that initiate discussions on genetics, biomedicine, and bioethics under the tutelage of experts (http://www.gesundheitsforschung-bmbf.de/de/186.php [14.11.09]).

[22] The first big citizens' conference dealing with genetics in Germany was held in Dresden in 2001 and discussed genetic diagnostics. Two years later, in the winter of 2003/2004, a citizens' conference on stem-cell research took place in Berlin. See Schicktanz and Naumann (2003), Tannert and Wiedemann (2004).

[23] The German Federal Ministry of Education and Research launched a "Science-Life-mobile" in 2000 that toured the country in order to enhance public knowledge of the factual issues in the life sciences. In 2003, it became the mobile "Bio-lab" cruising to schools in Southern Germany.

research in particular, predicts the Federal Ministry of Education and Research, will have far-reaching effects on our entire social life. To prepare the population for these pervasive upheavals, the Ministry is not only funding genetic research, but also genetic education. Accompanying research in the social sciences and the discourse between science and society should contribute toward producing a well-informed public.[24] The goal, according to the Federal Ministry, is to base decision making on comprehensible facts and rational grounds.[25] In the end this means the debate is shaped by expert-defined concepts and problems, not the common knowledge of citizens. Even at participative events purporting to democratize technology politics, citizens are not able to express themselves freely. Scientific authorities determine the subject matter, and discourse experts provide the framework in which citizens are to develop questions, answers, and their own opinions (Kerr, 2004, pp. 123–242; Niewöhner, 2004).[26] This framework leaves no room for the

---

[24] See the announcement of the Ministry in summer 2006 (http://www.gesundheitsforschung-bmbf.de/de/1276.php [5.8.2014]): progress in the life sciences and especially in human genome research and molecular medicine, the Ministry claims, poses serious ethical, legal, and social questions. Thus, it continues, in an open and increasingly knowledge based society, the critical discussion of these questions cannot be reserved to a small circle of experts, but has to be shaped and supported by a well-informed public, see also http://www.bmbf.de/de/1237.php [5.8.2014]. The Ministry not only funds discourse projects, but also the initiative "Science in dialogue" (http://www.wissenschaft-im-dialog.de/ [5.8.2014]) that organizes events and training in science communication.

[25] In a former version, the Ministry complains that public discussions about chances and risks and about ethical boundaries are emotionally charged, while it is necessary to base decisions on transparent facts and rational argumentation (http://www.bmbf.de/de/1056.php [2.12.09]).

[26] The political scientists Kathrin Braun, Svea Hermann, and others argue that public bioethical discussions mainly aim at imposing a specific way of thinking and speaking about people. The programmes and experts do not determine decisions and so-called "value judgments", but demand reflections based on the experts' worldview, that is on "scientific facts" and the belief in technological solutions, see Braun et al. (2008). Hermann (2009) has investigated the social function of bioethics: by reshaping

discussion of basic social issues such as the power of experts or the downside of technological and scientific progress. Instead, these issues are reframed into defined risks that can be balanced, reduced, and managed (Robins, 2001). Citizen participation stipulates faith in science: "There is an assumption that science is an important force for human improvement and that it offers a uniquely privileged view of the everyday world... Finally, science is portrayed in these accounts as if it were a value-free and neutral activity" (Irwin and Wynne, 1996, p. 6). The environmentalist who abhors genetic manipulation out of veneration for nature; the farmer who rages at the machinations of the agro-industry; and the concerned mothers who find genetically modified crops on the field simply sinister—they are all to become relics of the past. No longer shall thinking and actions be determined by intuitive antipathy toward manipulated tomatoes, scepticism toward feasibility claims, and mistrust of experts, but rather by what scientists disclose as "facts" and "opportunities and risks". No longer shall colloquial language and common sense be the foundation of democratic debates, but instead what experts declare to be factual and rational.

Industry is likewise committed to genetic education. Bayer, Schering, and Roche all want rational decision making citizens and informed consumers. We know that the greatest developments are worth nothing when people do not understand them and thus are not prepared to accept them, explained the chairman of the Association of Chemical Industry (VCI) in North Rhine in 2001, and sent off a mobile genetic laboratory to promote proper understanding among citizens (Minwegen, 2003). "Science needs the support of society. A well-informed public aware of the opportunities and risks of biotechnology is con-

---

social and cultural impacts of the life sciences as bioethical problems that are individually assessed, the right to informed choice finally appears as the only adequate solution. In this way, bioethics depoliticizes the dispute.

sidered a competitive edge in global competition. The public involved in developing the agenda of European research policy is a clear vision for the future" (En Route to the Knowledge-Based Bio-Economy, 2007, p. 5).

Youth in particular are targeted by genetic awareness campaigns.[27] In student laboratories and discourse projects they practise manipulating genes as well as ponder bioethical dilemmas of opportunities and risks. Mostly, these events are organized as "Future Workshops" that have to room for fundamental questioning, but ask participants to discuss the "how", the social conditions in which the technology should be implemented.[28] Between Flensburg and Munich, research institutes, museums, and chemical corporations have set up 50 public laboratories which offer "hands-on genetic technology". The mobile gene laboratory "Bio-Lab" has been touring through Southern Germany since 2003. Every few days it travels to another school to help familiarize middle school and high school students with gene technology.[29] Some secondary schools have already set up their own gene technology laboratories in which students themselves can tinker around with genetic material.

The goal of this campaign, sponsored mainly by industry, is clear: to make the "fascinating and forward-looking biosciences" more accessible to youth (Gläsernes Labor, n.d. a). To this end, teenagers are to be convinced of the ubiquity of genes

---

[27] The effort to activate youth is not restricted to Germany; in all countries, geneticists aim at raising genetic literacy and stimulating enthusiasm for genetics and gene technology, see, among others, the project "Let's talk about genes" in the UK; *Ecancermedicalscience* (2014, 8, p. 408; doi: 10.3332/ecancer.2014.408); and American Society for Human Genetics (2004).

[28] Participants in a pupils' conference in Bremen, for example, were asked to solve the financing problem of a geneticized medicine.

[29] The mobile "Bio-Lab" is financed by the industry and through public funds. The advertising agency "Flad & Flad Communication Group" organized the rolling laboratories and coached the experts who staff them.

and their technical manageability: thus they can, as an experiment description promises, experience "in an impressive manner" how "gene switches" work by transferring a jellyfish gene to a bacteria.[30] Or they can isolate DNA from tomatoes in order to see that all food contains genes. Every day, the students are told, they eat "genes": "Every day we consume about 1–2 grams of this carrier substance of genetic information — the 'blueprint of life'" (Gläsernes Labor, n.d. b). The youth can also "discover their own selves", a handout casually promises, by creating a genetic fingerprint[31] from their own cells.

Popular science is a constitutive part of scientific concept formation. Scientists not only promote their research to the public by giving it social relevance, but they themselves strongly believe in its significance. The final aim of popular science, Ludwik Fleck argues, is to instill a *Weltanschauung*. As the above quotes from the handouts show, in such laboratory courses, students are inevitably taught not only biochemistry and how to work with pipettes and reagents, but also the genetic worldview. A laboratory instructor delivers an impressive lesson in genetic perception to a group of tenth graders in the Life Science Learning Lab "Glass Laboratory" in Berlin-Buch. As he explains the biological bases of the genetic fingerprint, he points out that 98 percent of their DNA overlaps with *Pan troglodytes*, the chimpanzee. The geneticist quickly anticipates the obvious conclusion that in view of the considerable difference between a

---

[30] The description promises that the experiment will demonstrate how genes are transferred and turned on: you transfer the gene for a green fluorescing protein of a purple-striped jelly into e-coli bacteria. As a result, they give light, too. See http://www.genlabor-schule.de/cgi-bin/s_357.cgi [30.10.2009].

[31] The genetic fingerprint is a DNA profile used for the identification of individuals. Geneticists analyse the lengths of different highly variable DNA segments and "individualize" the resulting profile by relating it to a population. The random match probability, that is the probability that the profile of the sample (e.g. at a scene of crime) and of the tested subject coincide by accident, has to be made extremely low. M'Charek (2000) has analysed the construction of reference populations for this purpose.

human and a chimpanzee DNA cannot be that important. He teaches his young listeners that just one genetic modification is responsible for the fact that humans can speak and not chimpanzees. If chimpanzees possessed this gene, he speculates, then they could also learn to speak. The students who looked bored while pipetting at their workbenches suddenly perk up for the first time that morning. One of them promptly asks if this gene could not simply be transferred to chimpanzees. Now a fantastic, but fully serious, discussion unfolds between the students and the teacher about whether a gene transplant could enable chimpanzees to speak. The gene-believing expert affirms: yes, it's feasible, but it's not possible owing to "ethical considerations". However, technical obstacles still exist: for speech to be possible certain brain functions and organs such as vocal cords are necessary. Besides, he points out, such a gene transfer would be "complicated". During the lab work the tenth graders did not seem smitten by chemical formulas and experiment instructions. On the other hand, this lesson in genetic self-perception at the end of the practical training hits the bullseye. According to the message imparted by the genetic expert, humans are nothing more than genetic mutations of chimpanzees. The idea of making a chimpanzee speak by means of a gene transfer is what engages these students more than anything else this morning, and they talk excitedly about it.[32]

2.2.3. Genetic counselling

The goal of genetic education is to teach citizens to base their thoughts and actions on research findings from the genetic laboratory. They are to realize that DNA structure, Mendelian rules, and risk calculations are not just specialized knowledge for experts or material for biology exams, but also play a significant role in our everyday lives. They are called upon, as citizens and consumers, to make "informed choices" — in other

---

[32] I was able to observe, as a participant, the laboratory training in the Glass Laboratory in Berlin-Buch in October 2004.

words, decisions based on "genetic knowledge". Citizens are no longer to think and act according to their own estimation, but are to learn to base their deliberations and decision making on "facts and rational grounds", as stated by the Federal Ministry of Education and Research.

Those who learn about genetics by attending congresses or by visiting the Glass Laboratory are free afterwards to go home and return to their everyday lives. Nobody will be asked to act on anything immediately, and those who cannot make head nor tail of "genes", "risks", and "ethical dilemmas" can happily cast them aside. There is, however, a form of genetic education which has direct personal relevance and which requires citizens to act: genetic counselling. Genetic counselling is a professional service that prepares people to make a concrete choice — typically an "informed choice" about whether to undergo genetic testing or not. The explicit goal of the session is "medically competent, individual guidance in decision making", as formulated by the German professional organization of human geneticists (Kommission für Öffentlichkeitsarbeit und ethische Fragen der Gesellschaft für Humangenetik e.V., 1996, p. 129). In most countries, from Canada to the UK to Australia, genetic counsellors qualify for their occupation by training for this specific purpose. In Germany, however, genetic counselling is reserved for qualified medical experts, either full-fledged human geneticists or gynaecologists and paediatricians who, with a few years of continuing education, can acquire a qualification in "medical genetics". In a genetic counselling session, which typically lasts one to two hours, they explain DNA and chromosome structure, heredity rules, disease statistics, and genetic testing options. If the geneticist is sitting across from a pregnant woman who is to decide whether or not to undergo an amniocentesis,[33] he emphasizes the cellular processes involved

---

[33] Amniocentesis is a prenatal intervention for testing the fetal genetic make-up. By inserting a needle into the pregnant woman's belly, the physician samples amniotic fluid which contains fetal cells. Routinely,

in fertilization, statistics of birth defects, pregnancy risks, various prenatal testing options, and risk-related decision making. On the other hand, if sitting in front of him are men and women whose families have had multiple occurrences of breast or colon cancer, then the topics of genetic mutations, cancer statistics, and early detection usually take precedence. But regardless of the particulars of each individual counselling session, it is of great importance for genetic counsellors that their clients make an "informed choice". Several times during the session they make clear that it is the client's obligation to draw his or her own conclusions from the lessons about genes, risks, and testing options. Although it is called counselling, a geneticist tells his pregnant client at the outset of the session that he will give her no advice. She herself must figure out what she wants to do. He can only tell her that she must "seek advice from herself" (Samerski, 2002, p. 230).

In Germany, up to the enactment of the Genetic Diagnosis Act (GenDG) in 2010, slightly less than 50,000 genetic counselling sessions were held every year, with the counselling centres already operating at full capacity. However, considerably more prenatal chromosome tests were being conducted than genetic counselling sessions, that is to say, far more than 60,000. The number of molecular genetic tests performed was even higher: more than 200,000 in 2004 — with the trend on the rise. Clearly, far more people were being tested than counselled. As a result, human geneticists have declared a state of emergency. The field of genetic counselling is in desperate need of expansion, they say, in order to meet the "undersupply" of

---

chromosomes are checked for potential aberrations, the most frequent and famous being Down's syndrome. The procedure is usually done in the 16th week of pregnancy. Since it might induce an abortion (with a probability of 0.5%), amniocentesis is not used for screening all pregnant women. Next to amniocentesis, there are two more methods for prenatal genetic testing: (1) chorionic villus sampling, which can be performed earlier in pregnancy and delivers quicker results, but carries a higher risk of miscarriage; and (2) umbilical cord puncture, which is considered a risky procedure and therefore is not offered as a routine test.

counselling.³⁴ The new law mandates genetic counselling before and after prenatal or predictive genetic testing and thus again has increased demand. Thus, in order to fill the gap, lawmakers have offered physicians a crash course for genetic counselling in their field. Nevertheless, human geneticists assume that the actual demand for genetic counselling is currently not met in Germany, neither "quantitatively" nor "qualitatively". Considering this urgency, the establishment of genetic counsellors as an independent profession following the Anglo-American example is under discussion (Schmidtke and Rüping, 2013).

Genetic counselling is a paradigmatic example of genetic education. It is an intensive form of the genetic decision making instruction that representatives from science, politics, and industry are demanding for society: citizens are to be empowered to make "informed choices" via instruction on genes, risks, and testing options. Only then will they be considered competent and self-determined. Genetic counselling, according to a book on the ethical issues of genetic counselling, serves to strengthen autonomy and decision making competence (Hirschberg and Frewer, 2009, p. 9). Geneticists themselves also name autonomy as the most important goal of their service: in a discourse project Austrian and German genetic counsellors made it clear that for them the guiding principle of genetic counselling is the autonomy of advice-seeking clients (Griessler, Littig and Pichelstorfer, 2009, p. 287). The geneticists point out, however, that autonomy does not mean that clients should think and act as they deem best. Most citizens, they declare, are not at all capable of "autonomy". Genetic counsellors have to first "actively promote" it (Griessler, Littig and Pichelstorfer, 2009, p. 291).

---

34 In 2005, human geneticist Claus Bartram calculated a demand of 120,000 genetic counselling sessions yearly, see http://www.gfhev.de/de/presse/pressemitteilungen/GfH2005_Statement_Bartram.pdf [15.8.2014].

2.3. On the History of Genetic Counselling: Genetics as the Foundation of Sociopolitics

2.3.1. *The scientific management of hereditary dispositions*

The endeavour to establish a social order based on "factual knowledge" from the genetic laboratory is not new. From the beginning most geneticists saw themselves not merely as specialists in pea breeding and the crossbreeding of flies but as experts on the social question. By the end of the nineteenth century the first life scientists, researchers such as Jacques Loeb (1859–1924), Francis Galton, and Ernst Haeckel (1834–1919), were already dreaming of an artificial social order based on findings from the biological research laboratory.

Mendel's treatise was still slumbering unnoticed on the shelves when statistician Galton defined the term eugenics[35] (Galton, 1883) and championed the improvement of the human race by means of "breeding". The rediscovery of Mendel at the turn of the century launched the experimental science of heredity and seemed to push within reach dreams of the expert-directed improvement of society and the human race. In many countries, from Brazil to the USA and Europe, an influential eugenics movement developed and gained support. Its mission was to communicate to society the new principles of biology: Darwin's "Survival of the Fittest" and Mendel's heredity laws.

> [Eugenics] was a paragon of rational scientific planning applied to the management of the most hitherto unregulated aspect of human life, the reproductive process itself. Eugenics was dedicated to the preservation of the most important of all human resources, the germ plasm of future generations. Eugenics was thus based on the knowledge of a special class of experts, those trained in reproductive biology and genetics. It was rational planning par excellence. (Allen, 1997, p. 83)

---

[35] "Eugenics is the science which deals with all influences that improve the inborn qualities of a race; also with those that develop them to the utmost advantage" (Galton, 1904, p. 1).

The new reproductive experts imputed a pathological genetic disposition to anyone who did not conform to the norms of industrialized society—whether alcoholics, prostitutes, the unemployed, criminals, the mentally deficient, the physically deformed, or seditious trade unionists—and worked to eliminate them.[36] The promotion of self-determination and "informed choice" is obviously not what geneticists in the first half of the twentieth century had in mind. They did not see their fellow humans as decision makers in need of scientific education but instead as mere gene carriers who needed to be governed by experts. People were nothing more than transit stations of cultishly idealized genetic material—genetic material which, in their estimation, had been crossbred much too unsystematically and needed to be managed by scientific planning if they were to save the human gene pool. Counselling was among the measures through which a biologically based social order would be established. Consequently, the first heredity counselling centres, many of which opened their doors in the 1920s, had mainly eugenic goals: to promote the genetic improvement of the population.

In 1927 the Association of Marriage Counselling Centres documented the existence of 100 such centres in Germany. Couples wanting to marry were advised to base their decision as to whether to start a family on the quality of their genetic dispositions. Hereditary physicians analysed the family pedigree, speculated on the basis of crippled fathers or feebleminded aunts about "defective genes",[37] applied Mendel's rules and, if necessary, advised the couple against marriage and childbearing. The demand for such hereditary predictions, however,

---

[36] Diane Paul (1995) and Daniel Kevles (1985) have examined the history of eugenics in the Anglo-American world.

[37] In Germany, the term "gene" did not spread into medical and eugenic discourse for decades. German geneticists preferred to talk about "Erbanlagen", that is "hereditary dispositions". Even today, the less technical term "Erbanlagen" is used synonymously with "genes" in genetic counselling sessions.

remained low. Even in National Socialism, where marriage counselling and marriage fitness certificates were compulsory, hardly anyone sought out a "counselling centre for the cultivation of race and genetics"[38] of their own free will (Czarnowski, 1991; Soden, 1988).

2.3.2. More effective than coercion: Education and responsibility

Repression and paternalism in democratic societies are limited in their effectiveness; they always encounter resistance and opposition. Much more effective than social discipline through coercion is thus "governing through freedom", as the liberal governing principle is aptly named. People are not coerced to conform, but motivated through incentives to conform by choice.[39] This form of exercising power, which is described by Michel Foucault in particular, is much more subtle and permeates deeper than coercive governance. It aims not so much at the external but first and foremost at the internal. It governs not only actions but especially thoughts and desires.[40]

The first eugenicists relied on coercive measures as much as on education and guided voluntarism. In the 1920s and 1930s educational brochures, magazines, museums, and exhibitions evoked the power of heredity. Exhibits on heredity in the large health exhibitions such as GeSoLei (health, social services, and physical fitness exhibition, 1926) and the hygiene exhibitions in Dresden (1911 and 1930) were a given. The millions of visitors to these exhibitions were instructed not only in the anatomy of

---

[38] In German: "Beratungsstelle für Erb- und Rassenpflege."
[39] One way to make people adapt without constraint is to impute needs to them, such as the need for education, or counselling. The needy person will then willingly seek the corresponding profession service. Marianne Gronemeyer (1988) and Ivan Illich (1992) have written about the power of needs.
[40] Foucault derives this governance technology that targets interiority from pastoral power and the spiritual control technologies of the church, see, among others, Foucault (1990a; 1982).

the body, in the improvement of their own fitness, and in population statistics, but were also taught about the power of hereditary disposition and the successes of genetics. By viewing the family pedigrees of "asocial families", visitors were to be convinced that theft and unemployment were caused by genetic influence; and with a display on Mendelian inheritance they could test the regularity by which such hereditary dispositions are transmitted. Such popularizing events, however, had an entirely different goal than today's genetic literacy movement. By no means were citizens to learn to become experts themselves in order to be empowered to make informed and autonomous decisions. On the contrary, scientific findings were put on display mainly to impress laypeople and to defend the new power wielded by experts in management and politics. The exhibitions served to manifest and widen the gap between laypeople and experts (Nikolow and Schirrmacher, 2007, p. 16). Thus genetic education and popularization also had a different social function than it does today. The point back then was to prepare citizens for a genetically based social order; they needed to be convinced of the authority of the new experts and of the necessity for a eugenic population policy.

By the 1930s, most geneticists declared the scientifically supported racism of the first eugenic movement and its all too simplistic belief in genetic determinism to be scientifically obsolete. But by no means, however, did they abandon their vision of establishing a social order based on the findings of genetics. At the end of the 1930s the American businessman and philanthropist Frederick Osborne stepped forward to radically reform the eugenic project. He dreamed of a voluntary, enlightened eugenics without compulsory measures and racist discrimination. He wanted to reorientate the project of improving the human race through genetics to the principles of modern democratic societies. Citizens were no longer to be subdued, but instead taught to think eugenically on their own. By means of scientific education he wanted to instil in parents a sense of responsibility for the genetic endowment of their offspring (Wess, 1989, pp. 46–8). Soon afterwards, in 1947, the geneticist

Sheldon Reed coined the term "genetic counselling" (Reed, 1974). With this renaming he hoped to lift heredity counselling out of the discredited sphere of eugenics. No longer did he want to improve the national gene pool but instead the health of individual families; no longer did he want to pursue a heredity population policy, but instead hoped to motivate couples to follow their own interest in basing their family planning on genetic constitutions and heredity laws. A good twenty-five years later he assumed that this move to conceptually divorce heredity consultation from the shadow of eugenics ultimately helped the field of genetic counselling to thrive.[41] In 1951 there were only ten genetic counselling centres in the USA; in 1974 the number had already risen to 387.

In Germany human genetics was long held in contempt, and it was not able to redeem itself to the public until the 1970s. In Nazi Germany eugenicists, physicians, politicians, and bureaucrats implemented what had been previously conceived by geneticists at laboratory benches and desks. In Germany the findings on fruit flies elaborated by geneticists from different countries were not only theoretically but practically applied to human beings for the first time (Wess, 1989, p. 35).[42]

Human genetics became socially acceptable again when it received the seal of "healing art" by affiliating itself with medicine. A fundamental transformation in medicine, primarily in the 1950s and 1960s, was the prerequisite for this affiliation. Medicine changed from being a healing art that focuses on patients and their diseases to being a risk medicine that monitors the health of the entire population. The idea of "pre-

---

[41] "It is my impression that my practice of divorcing the two concepts of eugenics and genetic counselling contributed to the rapid growth of genetic counselling. Genetic counselling would have been rejected, in all probability, if it had been presented as a technique of eugenics" (Reed, 1974, pp. 335–6).

[42] Leading human geneticists of National Socialism such as Fritz Lenz and Otmar von Verschuer had been appointed to professorships and participated in international conferences soon after the war ended, see Weingart, Kroll and Bayertz (1992, pp. 562–81).

vention", that is, the prevention of diseases, rose to ascendency. Epidemiology, the statistical analysis of disease incidence and risk factors in populations, advanced to become the leading science. The focus of medicine was no longer the diagnosis and treatment of diseases but capturing and managing statistically calculated risks (Armstrong, 1995). In the 1960s Germany implemented the first prevention programmes for the healthy population, namely cancer screenings and the so-called "mother passport",[43] a pregnancy record intended to encourage women to undergo antenatal medical exams. In order to entice expectant mothers to see the doctor even when nothing was wrong with them, they were rewarded with 100 D-Marks when they completed all tests scheduled in the records.[44] This change in medical thinking made it possible for human genetics to slip under the umbrella of medicine. Medicine's main goal became prevention, and the main task of physicians risk attestation and risk management. A discipline dedicated to the classification and diagnosis of incurable anomalies could contribute to these new medical tasks. Henceforth, what before had been eugenics, namely the elimination of the unfit, once again became acceptable in the name of genetic risk management. The prevention, not of diseases, but of people classified as aberrant became a medical subfield.

In Germany the first genetic counselling centres opened their doors in the 1970s. Devoted to the guiding principle of pre-

---

[43] When my mother was pregnant with me in 1969, she got one of the first German "mother passports". It was a small document where the doctor noted the date of his haptic examination, fundal height, and the date of quickening according to the mother. My mother saw her gynaecologist six times. Her last visit was four weeks before my birth. In contrast, the "mother passports" for my daughters, born after 2000, are at least three times as large, provide for ultrasound scans and dozens of examinations, contain complex graphics about fetal growth, and list 56 potential risk factors that label the majority of healthy pregnant women as "at risk".

[44] The reward was scrapped only in 1983 after women had gotten used to the originally strange practice of visiting a doctor in order to have their health monitored.

vention, geneticists considered it their mission to keep people they classified as "abnormal", "disabled", or "diseased" from being born. If the risk they derived from empirical frequencies and heredity rules seemed too great for the prospective offspring, they advised the couple against marriage and childbearing. They hoped in this way to kill two birds with one stone: on the one hand, they believed this would save families from unhappiness and suffering, while welcoming on the other hand the eugenic side effect they anticipated from this "prevention of hereditary defective progeny".[45]

Most people, however, had no need for discouraging future predictions. After the first counselling centres opened in Marburg and Frankfurt am Main as state-financed model projects, human geneticists had to pull out all the stops to recruit clients. Those without a severe hereditary disease in the family simply did not see the need to consult a hereditary physician. Predictive health management was still a foreign concept to most people. Consequently, all preventive medical programmes lacked clients. In 1977, the human geneticist Ursel Theile complained that one of the unsolved problems in preventive medicine is the insufficient utilization of such opportunities by a still healthy population (Theile, 1977, p. IX). At that time those who were not ill saw no reason to visit a physician.

2.3.3. A new goal: The informed decision

During the 1970s and 1980s genetic counselling grew to include modern services for pregnant women and prospective parents. Risk and prevention thinking found its way into medical prenatal care, and new examination techniques, most notably ultra-

---

[45] Human geneticists did not have to give up their eugenic goals. They simply relabelled eugenics as genetic prevention. Human geneticist Gerhardt Wendt successfully promoted the re-establishment of genetic counselling in Germany, praising it as a means to protect society from an "afflux of the handicapped" and corresponding financial burdens, see Wendt (1978, p. 442). For eugenic arguments see also Baitsch (1970), Fuhrmann and Vogel (1975), Lenz and Lenz (1968).

sound, created a new patient: the unborn child. The fetus and its normal development became the focus of medical attention (Samerski, 2014). Ultrasound imaging and advances in chromosome preparation made it possible by the end of the 1960s to extract amniotic fluid through a hollow needle inserted into a pregnant woman's belly and thereby examine fetal cells for abnormal chromosomes. During the 1970s gynaecologists and human geneticists introduced chromosome testing by amniocentesis into prenatal care. Lawmakers relaxed restrictions on abortions and permitted abortions based on "eugenic indication".[46] Pregnant women undergoing amniocentesis at the recommendation of their physician suddenly found themselves in an entirely new state, in a state that sociologist Barbara Katz Rothman has called "the tentative pregnancy" (Rothman, 1986): they were no longer happily expecting a baby; now their whole pregnancy was on trial. Before they could actually expect a child, the test had to give them a green light. If the test showed evidence of any chromosomal abnormalities, they were urged to abort the pregnancy. Over time, genetic testing and the selection of future citizens has become commonplace.

The human genetic institutes received a big boost from these new selection technologies. Prenatal chromosome testing has provided them with a new source of income in the chromosome laboratory as well as new counselling clientele. Physicians suddenly started stamping "high-risk pregnancy" on entire pregnant cohorts and declared them "in need of counselling". The statistical probability of trisomy 21, the chromosomal finding for Down's syndrome,[47] correlates with the age of a

---

[46] In Germany, a new version of §218, the penal law regulating abortion, came into force in 1976. It ruled that abortion was exempt from punishment in cases of specific indications, among them the so-called eugenic indication. The Federal Constitutional Court dismissed a law permitting a first trimester abortion in 1975.

[47] In the era of prenatal diagnostics, Down's syndrome has advanced to the embodiment of the handicapped child that is preventable and decidable. Routine prenatal testing is mainly performed to track down developing children with Down's syndrome and to question their birth. It is

pregnant woman. Physicians first ascribed the rise in probability to all women over the age of 38 as an "age risk", and later, when laboratory capacity increased, to all women over 35. In doing so, they turned an entire population of healthy pregnant women into patients in need of special medical treatment, namely of prenatal chromosome tests. In the 1980s the first genetic tests for monogenic hereditary diseases were finally developed, in which the unborn child is classified not only as a chromosome carrier but as a gene carrier.[48] The days in which a geneticist could only engage in genetic marriage counselling were over. Finally they had something much more promising to offer than the unpopular recommendation, based on statistical frequencies and Mendel's laws, to break off an engagement or

---

estimated that of the pregnant women who are confronted with the test result "trisomy 21", the chromosomal findings for Down's syndrome, more than 90% terminate their pregnancy. Yet, there is a gulf between the laboratory findings and the child. "There is no way to 'be prepared' when given a diagnosis of Down's syndrome precisely because the diagnosis doesn't tell much about what the person might be. In other words, the diagnosis may be clear—see that third 21st chromosome and you can be sure that the fetus will have Down's syndrome. But the prognosis—what will become of this fetus—is not at all clear. Women contemplating prenatal diagnosis often say that they would terminate the pregnancy for severe mental retardation, but not for mild retardation. What you cannot tell looking at the chromosomes is how severe the retardation will be" (Rothman, 1998, p. 187).

[48] In the beginning, geneticists were able to perform so-called indirect genetic tests with the help of genetic markers, resulting in the assertion that the unborn, with a certain probability, had inherited the part of the chromosome with the mutated gene. After the discovery of so-called "genes for" such as the "gene for cystic fibrosis" (1989) or the "gene for Huntington's chorea", direct testing that searches for mutations at the corresponding DNA-sequences became possible. Even in the case of a so-called monogenetic diseases such as cystic fibrosis, however, geneticists cannot predict the disease from a genetic mutation: "No predictions can be made about the occurrence of common complications of cystic fibrosis or the severity or course of pulmonary disease, because of the wide variability in each group of patients carrying the cystic fibrosis genotypes studied" (The Cystic Fibrosis Genotype-Phenotype Consortium, 1993, p. 1311). Twelve years later, this conclusion was confirmed, see Slieker *et al.* (2005).

to forego childbearing. A product of its time, modern human genetics now offers "knowledge", "prevention", and a prenatal health check.

Prenatal quality control has now become routine. In recent years approximately every tenth German child born has already had its chromosome set tested by amniocentesis.[49] Practically every pregnant woman is examined several times with ultrasound imaging, and many women have their risk of a fetal chromosome disorder calculated with a first trimester screen. A new blood test[50] examining cell free fetal DNA from the mother's blood is on the verge of revolutionizing prenatal testing: it promises a safe genetic check-up for a growing number of conditions without risk of intervention. After a simple prick, geneticists can, theoretically, check the genetic make-up of the unborn child for aberrations. As an additional prenatal screen for chromosomal disorders, it is already on the market in many countries. If the test is technically upgraded and becomes routine, the pressure on pregnant women to submit to testing will be tremendous. For what reason can a responsible expectant mother refuse a risk-free health check for her baby?

This intensive medical monitoring of a healthy population is paradigmatic of the new form of risk-orientated medicine, which the medical sociologist David Armstrong calls "surveillance medicine" (Armstrong, 1995). Surveillance medicine, according to Armstrong, differs considerably from the previous clinical medicine. First of all, it blurs the difference between "normal" and "pathological", which had hitherto shaped medical thinking and acting. Prenatal diagnostics declares that

---

[49] Numbers taken from Schmidtke and Pabst (2007).
[50] The new "blood test" measures fetal DNA in the pregnant women's blood to check for common numerical chromosomal aberrations. It produces false negatives and false positives and therefore is classified not as a diagnostic test but as a screening test; positive results have to be confirmed by an amniocentesis or chorionic villus sampling. In the general population, the positive predictive value is less than 50%, that means a 30-year-old pregnant woman with a positive test result has a 50% probability of actually having a child with Down's syndrome.

all pregnant women are in need of care—even if nothing is wrong with them. No longer are the ill the sole targets of medical monitoring and treatment but the healthy population in particular: "Surveillance medicine requires the dissolution of the distinct clinical categories of healthy and ill as it attempts to bring everyone within its network of visibility" (Armstrong, 1995, p. 395). Second, prenatal diagnosis is no longer concerned with healing but with risk management. The aim is not to restore lost health but to take possession of the future. "Surveillance medicine... attempt[s] to transform the future by changing the health attitudes and health behaviours of the present" (Armstrong, 1995, p. 402).

This problematization of the healthy and anticipation of the future have had dramatic effects, especially in pregnancy. Prenatal testing creates patients who cannot be helped. They cannot be healed, only aborted. But who may make such a decision? Who may determine which diagnoses and risks are reasons to prevent people from being born? In Germany until 1995 a pregnancy termination based on prenatal diagnosis fell under the so-called eugenic or embryopathic indication. This indication suggests that there could be medical reasons for preventing the birth of exceptional people. In fact, until the 1980s it was customary for physicians and geneticists to judge whether a child should be born or not in view of its predicted developmental odds. With the revision of the German Abortion Law (§218) in 1995, however, eugenic indications have become a medical indication in which a pregnancy can be terminated at any stage without penalty if it presents, as it is formulated, a danger to the life or the risk of serious impairment of the physical or mental health of the mother. Consequently, the law no longer justifies abortion because of an abnormality in the health of the unborn child but because of the anticipated burden for the prospective mother. Although the indication must be determined by a physician, the law hands the decision over to the pregnant woman. She has to sign a paper stating that life with an ill or disabled child would be unacceptable in her circumstances. Based on prenatal test results, she is to imagine life with her future child and then decide if she still wants to bear it. A physician is to guide her in these deliberations.

Women are asked to make their own choices, but only after they have been instructed by professional counselling. The German Abortion Law obliges women who want a regular first trimester abortion to undergo "counselling of pregnant women in conflict situations" at pregnancy counselling centres. Lawmakers define the desire to interrupt pregnancy as a conflict with the pregnancy. Counselling has to be an open outcome process, the law states, but has the explicit purpose to protect the unborn child. And the new German Pregnancy Conflict Act of 2010 obliges women to be counselled by a physician before they can undergo a selective abortion. This is to ensure that they do not simply act as they want but instead make an "adequate" and "balanced" decision (Bundesärztekammer und Deutsche Gesellschaft für Gynäkologie und Geburtshilfe e.V., 2009). According to body historian Barbara Duden, this compulsory counselling can only be understood as a subtle form of perception management. It obliges women to experience the amalgamation of two heterogeneous realities: pregnancy as a tangible culturally experienced state on the one hand and pregnancy as a sense-less, diagnosed fact on the other hand (Duden, 1996, p. 96).

Not only concerning abortion, but also concerning prenatal tests, the German law has established compulsory counselling. Since February 2010 the new Genetic Diagnosis Act (GenDG) makes decision counselling with a human geneticist compulsory before prenatal or predictive genetic testing. Here again lawmakers want to ensure that the pregnant woman will make an "informed choice". In practice, however, compulsory counselling has existed since the 1980s. German law has gradually imposed a physician's duty of counselling for prenatal diagnosis—and not only for women who want to be tested but prospectively for all pregnant women: if a gynaecologist does not explicitly inform his pregnant patients of the probability of a genetic abnormality and the relevant prenatal testing options, he or she runs the risk of bearing the responsibility for the life of a disabled or ill child. On several occasions gynaecologists and geneticists have been sentenced to lifelong payments of child support because women—respectively their lawyers—were able to credibly show that their inadequate instructions were responsible for the birth of their child.[51] The child was only born, the women argued, because they were not made sufficiently aware of pregnancy risks and prenatal tests. Since then physicians have had to inform their patients about possible deformities, risks, and prenatal check-ups. Thus, before they notice their belly swelling or feel the first stir from the baby, pregnant women are already informed about everything that can go wrong, what risks they are taking, and what decisions they have to make.

Physicians are legally on the safe side when they recommend prenatal testing for their patients. This may be a reason why many gynaecologists ignore their own professional guidelines which state that women are to decide for themselves. They urge their patients to undergo amniocentesis and ultimately also to terminate the pregnancy (Braun, 2006). Unlike gynaecologists, on the other hand, geneticists take great care to hand over all decision making to their clients. They explicitly uphold the counselling maxim of not offering any recommendations. They see their task as facilitating clients' decision making. Genetic counsellors around the world are practically unanimous in their acceptance of this basic principle, in professional jargon called "non-directivity".[52] It is celebrated as an emancipatory

---

[51] The German legal system denies the possibility to issue a wrongful life claim because it is seen as unconstitutional. Thus, jurisdiction does not declare the handicapped child a damage, but the parents' costs of raising it. The Federal Court of Justice has confirmed corresponding court decisions that regard the full support for the child as a damage on 18 June 2002.

[52] The term "non-directivity" originates in the client centred counselling concept of psychotherapist Carl Rogers (1902-1987), where the counsellor does not preset the session but helps his client to clarify problems and find solutions on his own. Genetic counsellors have adopted the

achievement, as a renunciation of eugenic objectives, and as a bulwark against the state population policy. The landmark of this transition from the paradigm of prevention to the paradigm of decision making in Germany is a book titled *Genetic Counselling: Assistance in Making a Self-Responsible Decision?*, co-authored by human geneticist Helmut Baitsch and psychologist Maria Reif (Reif and Baitsch, 1986). The authors dismiss all eugenic and preventive ambitions and pronounce a new goal for genetic counselling: the self-responsible decision. Through a combination of scientific instruction and psychosocial support, geneticists are to empower their clients to be capable of making a self-responsible decision which they understand and can live with in the long-term (Reif and Baitsch, 1986, p. 13).

Today, pregnant women who are to be empowered to make decisions about prenatal testing options are still the most important clientele for genetic counselling.[53] On the rise with the number of predictive genetic tests is the number of women and men who must decide about a genetic check-up to predict their own future. The majority of such predictive tests forebode a frightening disease for healthy people—in the form of a genetic risk. Apart from rare exceptions like Huntington's disease, only probabilities can be derived from genetic findings, that is, statistical frequencies which are then ascribed to clients as the risk of contracting a disease. The test attests their elevated risk for cancer, hemachromatosis, Alzheimer's, or stroke. Since genetic research has committed itself to the search for risk genes or susceptibility genes, a greater number of genetic tests which categorize healthy people into a genetic risk profile are coming on the market. Years ago German human geneticists resolved to permit genetic testing only for those who have been counselled

---

term in order to label their principle not to give advice and recommendation.

[53] If a family already has a child with a disease or handicap that is diagnosed as genetic, women or couples visit a genetic counsellor before they become pregnant (again). These counselling sessions are not fundamentally different from the sessions with pregnant women.

(Berufsverband Medizinische Genetik e.V., 1996; Berufsverband Medizinische Genetik e.V., 1997). The Genetic Diagnosis Act (GenDG) now mandates genetic counselling before a predictive test. However, a growing number of private companies such as "DeCodegenetics" and the Google-sponsored "23andMe" are entering the market and offering genetic testing on the internet without regard to national laws. As a remedy for this, scientists and politicians have called for more counselling and education. To protect citizens from fraudulent claims for genetic testing, they want to train citizens to become educated consumers. They hope the average citizen can be empowered to find her way through the jungle of genetic promises, genetic testing options, and genetic risks. So that Google cannot lead her down the garden path she is to be taught to make informed decisions.

Seventy-five years after philanthropist Osborne dreamed of a society whose members had learned to think and act according to eugenic principles, genetic education has become commonplace. Today numerous informational and counselling events aim to encourage people to make decisions based on genetic information. In the following chapter I will examine more closely such decision making instruction. Taking the example of genetic counselling I will look at what geneticists teach citizens when they inform them about genetic errors, ominous risks, and genetic testing options. In these sessions, the symbolic power of genetic education is quite evident: a geneticist and laypeople sit down together and engage in a personal conversation. The geneticist addresses his explanation not to an anonymous public but to a specific person. Counselling clients thus expect that the expert is making statements that are comprehensible and meaningful to them. The geneticist, in turn, is striving to convey his expertise in a way that seems relevant to his client. He lends not only everyday importance to his knowledge but also applies it personally to his client. As a result, the example of genetic counselling very clearly reveals the form of thinking and acting demanded of citizens in genetically informed decision making.

I will first refer to my observations of counselling sessions in which women are being prepared to make an "informed decision" about genetic testing for breast or colon cancer. The genetic decision making lesson is more critical and charged when the attested risk and the test being offered question not the future health of a client but the bearing of her child. Therefore at the end of the third chapter I will expressly address a genetic counselling session in which a pregnant woman is turned into an informed decision maker about her coming child.

All the genetic counselling sessions I report about and quote from I have recorded on tape as a participant observer. Additionally, I made notes on nonverbal communication and occurrences. All quotations are taken verbatim from my transcripts (for transcript conventions, see appendix). The first four discussions about genetic testing for breast and colon cancer took place at a centre for cancer genetic counselling at a large university clinic. The genetic counselling sessions held with pregnant women were observed at the genetic counselling centre of a university women's clinic and at a university institute of human genetics.[54] At times I turn to excerpts and formulations from another corpus of counselling sessions when they seemed particularly telling or illustrative. These I have also observed and recorded at university genetic counselling centres, and subsequently transcribed.

---

[54] I have analysed the genetic counselling of pregnant women in my book *The Mathematization of Hope*, see Samerski (2002); in the course of my research, I also observed genetic counselling sessions on cancer genetics and recurrence risks of other diseases which I have partly used for the research that has resulted in this book. In Spring 2004, I observed and recorded four genetic counselling sessions on colon cancer and breast cancer at a major German counselling centre. I abstain from mentioning names not only for the protection of privacy: I do not want to criticize single counsellors or counselling centres. The absurdities and confusions I will describe in the third chapter of the book are characteristic of genetic counselling in general. It would be misleading to blame single persons or institutions.

Chapter Three

"Informed Choice"

How Genetic Counsellors Empower their Clients to Attain Self-Determination

Ms M. is sitting in the counselling room of a large university clinic. Sitting opposite her is a geneticist. They are in the middle of a counselling session. The geneticist has been asking Ms M. a comprehensive battery of questions about her health and the health of her relatives, and is busy making notes. The question period is now finished, and the geneticist changes her mode of speech: she switches over from questioning to lecturing. Now she will teach her client about genes, mutations, disease risks, and testing procedures: "Now I would like to talk to you a little about genetic testing, what genes are, what we mean by inherited, and so forth, okay?" Ms M. is in her mid-forties and was sent by her family physician to the genetic counselling centre. He recommended she see a human geneticist after she had a polyp removed from her colon and had informed him that some of her relatives had had colon cancer. Now she hopes the geneticist will tell her more about this disease, her own susceptibility, and precautionary measures. Several times during the session she states that she would like to actively undertake measures to ward off the disease. Her mother, she explains, had already warned her about the prevalence of colon cancer in the family: "Because my mother told me that everyone around her died of colon cancer", she says, adding, "And I don't want that for myself."

The geneticist regularly conducts genetic counselling sessions. She is a young physician training to specialize in human genetics. Her job is to teach people about genes, genetic risks, and genetic testing in order to enable them to make an informed decision. Professionally she has chosen to specialize in the area of chromosomes, DNA, and biostatistics, and it is from this perspective that she regards her client. That she will only inform her client and not offer advice is a counselling principle she has already internalized.

While instructing Ms M. she offers her a glimpse into the world of her expertise: she talks about DNA, base pairs, triggering genes, mutations, mismatch repairs, and cancer incidence rates. Then she informs her client that she meets the international criteria for an at-risk person. Hence there is an increased probability, concludes the geneticist, that Ms M. could have a gene mutation—whereby she strongly emphasizes the "could". Therefore she is eligible for genetic testing and must be diligent about complying with routine early detection measures. "As a precaution", she says. However, the test is only an offered service: it is entirely up to Ms M. to decide if she wants to undergo testing or not. Ms M. alone must decide, the geneticist repeatedly stresses, what the "correct path" is for herself.

The goal of the genetic counselling session is to promote Ms M.'s self-determination. She will learn to make an autonomous, self-responsible choice for or against screening, as geneticists put it (Zoll, 2009, p. 87).[1] To this end the next ninety minutes are spent teaching her about DNA, cancer risks, genetic analysis

---

[1] Genetic counselling is to assist an individual or a family to understand medical-genetic facts, to deliberate on decision making options, and to select an individually appropriate mode of behaviour—this is how the German Medical Genetics Professional Association describes the meaning and purpose of such sessions (Berufsverband Medizinische Genetik e.V., 1996, p. 1).

procedures, early cancer screening, and decision making options. What Ms M. ultimately does with these explanations is her affair. Particularly with regard to the offered genetic test, the geneticist refuses to draw any conclusions from her lesson. This is Ms M.'s obligation, she emphasizes. "I can convey information to you", says a colleague from another counselling centre, summarizing the principle of genetic decision making instruction, "and how you deal with it is ultimately up to you."

3.1. The Initial Transformation of the Person: The Client as a Gene Carrier

The most important input that the client of genetic counselling receives for her self-determination are lessons about genes and risks. She is told about the double helix, mutation possibilities, Mendelian heredity laws, and the probabilities of genetic disease occurrence. The goal of these lessons is to inform the client about herself. After all, she is not being prepared to take an exam but to make a decision concerning her health and her future. What she hears, therefore, claims to have direct relevance for her life. The genetics lesson thus inevitably reframes the client. Talk about "genes" inevitably raises the question: what is a person?[2]

3.1.1. The genetic person

The manner in which geneticists reframe their listeners and interlocutors when they teach them about genes is demonstrated by an example of a public discussion event in Hanover. The director of the Hanover Institute for Human Genetics, Jörg Schmidtke, has taken it upon himself to draw the public's attention to distorted notions about genes.[3] He is disappointed

---

[2] Barbara Duden named the gene in colloquial language the "reflexive gene", since it inevitably reframes the addressee and speaker into gene carriers. See Duden and Samerski (2007).

[3] Jörg Schmidtke's lecture was titled "The Unwavering Belief in the Power of Genes—A Critical Analysis from the Perspective of a Human Geneticist". This lecture was delivered at the conference *So What Exactly*

that so far they have not been successful in appropriately conveying "the role of genes in human life". So now he wants to teach his fellow citizens about genetics.[4]

Right at the beginning of his speech Schmidtke points out to his listeners that they are the object of his expertise: he will be talking about the "human species to which we all belong", he clarifies. Within this species there are only insignificant genetic differences, he goes on to explain. These sentences set the framework for the seminar. First he appropriates everyone in attendance, whether they like it or not, into an inclusive, global, and inescapable "we all" as the "human species". Then he appoints himself as an expert about this biological "we" by declaring that all people are gene carriers. He does not talk to his listeners as peers but as members of a biological species about which he possesses scientifically validated knowledge. Although the people in the audience are the targets of his explanations, they are also the objects of his specialized knowledge. This framework implies that objections or protestations of common sense have no place here. As a geneticist only he knows about gene carriers: in other words, about the people in the audience. A common ground for dialogue therefore does not exist.

As a psychogeneticist Schmidtke is searching for genes that control behaviour. To this end he is researching the mating behaviour of the female rhesus monkey. Hence he feels empowered to explain the causes of marital infidelity to his listeners — mostly women.

---

is Normal? Notions of Health, Illness, and Disability in Genetics and Society, held on 6–7 October 2005, in Hanover, Germany. The conference was organized by the Centre for Health Ethics at the Protestand Academy Loccum in cooperation with the IMEW (Institute for People, Ethics and Science) in Berlin. The quotations are verbatim translations from the notes I took during the German lecture.

[4] Schmidtke is not an exception with this desire. Since the gene hype of the 1990s geneticists have been increasingly stepping forward as educators who wish to free their fellow citizens from false beliefs about genetics. See here, among others, also Fraser (2001).

As he has already reframed being human as belonging to the biological species *Homo sapiens*, he can now talk about the limbic system, copulating female monkeys, and marital fidelity in a single breath. He knows—only a few genes, after all, separate female monkeys from women—that serotonin levels contribute to the occurrence of extramarital escapades by the female sex. As a champion of genes that do not determine but merely dispose, however, he does not want to simply excuse the unfaithful. "Genes", he explains, are "cross-linked information carriers" that "sometimes crash" and receive external "commands". Thus human beings are not victims of their genes. They can learn to live with their genes. The prerequisite for this, however, is the message of his presentation, namely that people should be enlightened by geneticists. Those who do not want their genes to give them the runaround must become aware of and be informed about their genetic inheritance and then actively choose their behaviour. Instead of giving in to the infidelity gene, suggests Schmidtke casually, one could eat chocolate instead. Chocolate contains a serotonin precursor.

Schmidtke is not an exception: human geneticists believe it is their vocation to teach their fellow citizens about themselves. They reframe people into two-legged gene carriers and make it clear to them that they need genetic education. As a bundle of DNA, mutations, hidden information units, and probabilistic gene effects they can no longer know themselves. Those who wish to be "self-determined" must go to a geneticist to find out what their "self" actually is. In the age of genetics, self-determination and autonomy no longer mean being without supervision; rather they presuppose having been taught about oneself by a genetic expert.

3.1.2. *The incomprehensible self*

Genetic counselling also reframes people as two-legged gene carriers. Ms M. receives a lesson about biochemistry and molecules together with information about herself. This instruction is still much more powerful than Jörg Schmidtke's proselytizing. Schmidtke delivered his lecture to an anonymous audience. By contrast, the genetic counsellor is engaged in a personal conversation with her client. What she is explaining is addressed directly to Ms M. with the goal of enabling her to make a decision. In so doing, she asks her client to think of her own internal body as a site of genetic modifications, mutations, and disease probabilities — and to base her decision making on this. Thereby the geneticist not only turns all people in general into a genetic case but also Ms M. in particular. What kind of new "self" the geneticist is assigning to her client, and what fears, needs, and illusions she is awakening, are questions I will now explore one by one on the basis of the genetic counselling session.

First of all, the geneticist begins with a general lesson about the inside of a cell nucleus.

For approximately twenty minutes she teaches her client about DNA structure, genes, chromosomes, possible gene mutations, and Mendelian inheritance. Most of the time she is teaching her client about abstract interrelations that she occasionally illustrates with visual aids and analogies. She opens a thick binder holding visual aids: schematic black-and-white diagrams of chromosomes, the double helix, and the genetic code. Then a standard lesson begins:

> C: Here is a simple illustration. This is what a chromosome looks like. Here are the various bands, and here are the genes lined up like beads on a string, interspersed by sequences that do not encode the genes themselves and thus proteins, but rather function merely as switches.

Ms M. admits that she knows very little about genetics. So the geneticist's presentation is not refreshing existing knowledge but is teaching her something new. But Ms M. cannot really understand what is told to her. For the most part she is being confronted with incomprehensible jargon. The geneticist is talking about "bands" and "sequences" without explaining what they mean to her client. Ms M. does not know that "bands" are the outcome of a certain colouring technique and that "sequences" refer to base sequences, that is, DNA segments. It is not possible to decipher such technical terms outside their original context.

After the counsellor explains that genes "lie" on the chromosomes, she enters into a detailed explanation of the genetic code and possible mutations. Again she places in front of Ms M. an illustration. This time we see a band composed of the letters ATCG. The geneticist explains:

> C: So, for example, when this thymine, represented here, is not a thymine but a cytosine instead, or if the base is missing or there is an extra one, then there can be errors in the sequence, uhm, which can, uhm, produce a modified protein molecule. Here's another picture of this (leafing through pages) that shows how... (-) this is then read and transcribed into proteins.
> W: Hm=hm.
> C: This is what ultimately, so to speak, results on a gene, on such a point. And, um, these modifications, that is, from base exchanges or inclusions or exclusions, can also lead to a breakdown so that the protein molecule is not formed at all. Then the associated function in the body van... um... is lost.
> W: Hm=hm.

Ms M., perplexed, looks at the counsellor. "Thymine", "base", "sequence", "base exchanges", "read", "transcribe" — her facial expression betrays that she is not comprehending much. Even if she has heard one or two words before, they cannot really tell her anything. The technical terms refer to contexts unfamiliar to her. She is not acquainted with the scientific edifice that imparts meaning to these terms. Geneticists, on the other hand, are members of the genetics thought collective. Words like "base" and "thymine" are technical terms defined by the tenets, research methods, as well as the conceptual and theoretical assumptions of this thought collective. Outside this context these terms are meaningless. Laypersons such as Ms M. do not understand what a "base exchange" is. Moreover, Ms M. is listening in a different way. Scientific interest has not motivated her participation in the counselling session, but rather her concern that she may develop cancer. She does not want to know about biochemical and molecular genetic models; she wants to know more about herself and what preventive measures she can take.

Obviously her counsellor is aware of this discrepancy between her scientific crash course and the everyday-orientated understanding of her listener. Later she apologizes for having "thrown so much information", as she says, at her client. Not once does she check to see if Ms M. can follow her and has understood her. She pours out her explanations as if she does not expect her client to understand anyway. Her sentences are strikingly fragmented and truncated. A few intimations are not even comprehensible to professionals. A listener with basic genetic knowledge can guess that with "breakdown" she is referring not to a breakdown of the DNA string or the protein molecule but rather to a breakdown of the transcription process. But what she means when she says "This is what ultimately, so to speak, results on a gene, on such a point" is not clear even to listeners well versed in genetics.

Ms M.'s geneticist is by no means unfocused or inept.[5] It happens frequently that genetic counsellors rattle off specialized knowledge and pack it in unintelligible, even confusing, sentences. In the following excerpt another geneticist explains to a pregnant woman the risk her child carries of being born with an inherited disease and thus which genetic tests are relevant.

> C: Today there are tests at this level of the genetic material, DNA we call it, deoxyribonucleic acid, that can detect gene mutations. And we call it molecular genetic testing, at this level of genetic material. It's obtained by drawing blood and then, um, its structure is analysed. It's a terribly complicated procedure and the structure of this genetic material is known according to, um, it's known that it is made of four building

---

[5] After lectures I often hear the reproach that I selected marginal counselling centres and unprofessional counsellors. These quotes, however, are not unusual but are typical of genetic counselling sessions. All counselling sessions from which I quote were held at high-ranking university institutes and university clinics.

blocks, four bases, we say, adenine, guanine, cytosine, and thymine. You don't need to remember this (writes), so I'll just abbreviate these four building blocks, and hydrogen bonds always link them AT, A with T, G with C, and there are 3 billion such base pairs, such a la, like a ladder, a double helix like this is often depicted. Uhm, they, their sequence practically encodes genetic information and how everything in you functions.

For an uninitiated listener it is impossible to comprehend or understand anything. The presentations are interspersed with jargon and concealed intimations. With the remark "you don't need to remember this" the geneticist indicates that she does not even expect her client to understand. In the last sentence, however, she stresses the importance of this knowledge that she can only intimate to her client but not make comprehensible. Everything she is presenting, she claims, encodes nothing less than "genetic information and how everything in you functions".

Even though clients cannot really understand such presentations, the genetics lesson still says something to them. They are receiving a demonstration that experts are in possession of meaningful knowledge to which they themselves have no access.[6] The geneticist is sharing specialized knowledge that her client cannot understand but should regard as meaningful knowledge about herself. Clients worried about cancer and pregnant women should think that such incomprehensible things as "bases", "breakdowns", "sequences", and lost functions occur inside them. Their destiny hangs on these things known only by experts. If Ms M. wants to know something meaningful about herself she must rely on second-hand

---

[6] Other geneticists involved with so-called scientific communication also consistently intersperse technical terms in their explanations with the obvious intention to impress. When Jörg Schmidtke, for example, during his lecture in Hanover, wanted to convince his audience that a single gene can have a huge effect on human appearance, he talked about it "triggering a differentiation cascade". In doing so he was merely informing his listeners that he is in possession of meaningful knowledge that is beyond the understanding of the laypeople in the audience.

information—information that she cannot verify and assess but must accept as authoritative knowledge.[7] Ms M. has an intangible inner life, is thus unknowable to herself. She has learned that what constitutes herself lies beyond her own horizon. The geneticist has thus shown Ms M. how unknowledgeable and powerless she is—with regard to herself.

3.1.3. Things in the body

Confusing and unintelligible monologues that serve primarily to make laypeople aware of their ignorance is routine in genetic counselling. It would be too easy to ascribe it to the didactic or rhetorical incompetence of the quoted geneticists. The genetic counsellor is not impossible but her task is: she must take scientific abstractions out of their context and make them the subject matter of a generally comprehensible explanation—and in a manner that enables her client to make an "informed choice". There are two ways she can deal with the gulf between her genetic expertise and the everyday-orientated understanding of her client: either she strives to be scientific and overwhelms her client with jargon and fragmented textbook knowledge, or she trivializes her specialized knowledge and conveys it in neat images and analogies. So far the geneticist has mainly overwhelmed Ms M. Very often, however, geneticists strive to avoid scientific lecturing and remain as intelligible as possible to laypeople. They omit complex details and contexts that can only be explained with technical terms. Using comparisons, images, and colloquial paraphrasing they attempt to elucidate such scientific constructs as "gene", "mutation", and "risk". In the process, what science regards as abstract, hypothetical, and only conditionally valid is turned into a scientific fact. Following Ludwik Fleck, geneticists who do this are conveying to their clients not science but popular science—a specific form of knowledge: "Simplified, lucid and apodictic science—these are

---

[7] At best she can replace it with other second-hand authoritative knowledge, that is, with other expert opinions.

the most important characteristics of exoteric knowledge. In place of the specific constraint of thought by any proof, which can be found only with great effort, a vivid picture is created through simplification and valuation" (Fleck, 1981, pp. 112–3).

The genetics explanation is based on the two central concepts of "gene" and "risk". Yet rarely does the geneticist actually explain them. I have observed three dozen counselling sessions with altogether six different geneticists at four genetic counselling centres, and in not one session did a geneticist talk about the specific premises and the validity of statistical probabilities (see chapter 3.2). Nor is the gene, for nearly a century the basic building block of genetics, explained in most sessions. At one counselling centre the geneticists routinely confined themselves to equating genes with "Erbanlagen" and assumed there was no need for any further explanation. "Erbanlagen" is a common German term meaning "hereditary dispositions" or "hereditary factors". Most other geneticists trusted in their clients' general knowledge. They spoke about "genes" as if it were completely clear what was meant by the term. Unlike Jörg Schmidtke, who complained about distorted notions of genes in the general public, his colleagues assume during a counselling session that the term "gene" does not need to be explained. Only one geneticist broke ranks and tried to convey the idea of "genes" to his clients. He explained to them that there was a library of administrative directives in their cell nucleus. Standing on the shelves, for example, was "Volume Ear" or "Volume Nose" (Samerski, 2002, p. 156).

Most of the genetics education was based on undefined and seemingly universally intelligible scientific terms. No matter whether the geneticists were explaining hereditary laws, risks of deformity, chromosomes, mutations, disease probabilities, or genetic testing—the "gene" was more or less always explicitly underlying their explanations. The chromosome was described, for instance, as the carrier of genes; heredity as the transmission of genes; and disease probabilities as the outcome of "gene defects". And although—or perhaps because—they did not explain "gene", it takes on a peculiar form in their presentations:

the gene becomes a little thing with all kinds of impressive powers. The hypothetical nature of the technical term "gene", its specific premises and unclear reference, is all lost in the counselling session. Illustrations and colloquial formulations give the controversial, even antiquated, laboratory construct the appearance of being a scientific fact. How the reification of the gene in genetic counselling takes place, and what the clients learn from this, will be examined in the following sections.

3.1.3.1. Visual representations as reproductions of reality

Like all genetic counsellors, Ms M.'s genetic counsellor also enlists the help of visual aids and invites her client to take a look at the "scientific facts", as she explains. What cannot be grasped in generally intelligible words is supposed to become apparent through visual representations. Ms M. should see with her own eyes what is not even visible: "This is what a chromosome looks like", comments the counsellor about one chart, and "Here are the genes lined up like beads on a string".

Visual representations have an important function in science. In the generation and development of scientific facts (Fleck, 1981) they serve as intellectual crutches, as heuristic aids, to illustrate and clarify hypotheses and theories. Thus they suggest a new idea to a thought collective, saying: "We could look at it this way." The image of the genes lined up like beads on a string on the chromosome, for example, dates back to the work of fruit fly geneticist and Nobel Prize laureate Thomas Hunt Morgan. Long obsolete in genetics, the image still prevails in popular scientific representations. Morgan underpinned his fruit fly experiments with the assumption that genes are linearly arranged on chromosomes, and had tremendous scientific success (Rheinberger and Müller-Wille, 2009). Gradually this hypothesis of the bead-like arrangement of genes congealed into a scientific fact and took up residence in the model of "beads on a string". Such an image, intended to render a concept comprehensible, objectifies "in more ways than one: it gives something imagined the status of a concrete object and thus is suggestive of greater 'objectivity'" (Pörksen, 1997, p. 137). Although first research findings contradicted this model back in 1940s,[8] the "beads on a string" image nonetheless entered textbooks and has been consolidated into a canonical image. This changed the function and proposition of the image: as an established scientific fact it no longer proposes but confirms with great certainty: "This is how we look at it." Textbook illustrations and canonized images no longer serve to open new paths of thinking but rather reinforce and transmit existing ones. Their purpose is that of transmission and safeguarding for a knowledge community and integration into an edifice of teachings (Pörksen, 1997, p. 133). In genetic counselling, however, where esoteric knowledge is supposed to be translated into exoteric knowledge, such visual representations have yet another purpose: they serve neither to propose new ideas nor to promote better comprehension nor to introduce novices to the edifice of teachings. Ms M. is not being familiarized with a new scientific idea or being initiated into a subject; rather she is to consider what the geneticist is telling her as important knowledge about herself—knowledge on which she should base a momentous, personal decision. Here the visual representations purport to represent reality, thereby transforming scientific constructs into seemingly concrete objects. The image of the bead of strings or the double helix placed by the geneticist on the table claims: "This is how it is." This is how it looks inside cells. It is a representation. And not only of what is in any old cell but what is inside Ms M. She is not only the addressee of the genetics lesson but also the putative object. The representation says not only "This is how it is" but also "Look, this is you". Thus it purports not only to represent any reality but specifically the personal, physical reality of Ms M.[9]

---

[8] For example, the discovery of the so-called position effects, that is, of the different gene effects depending upon its position on the chromosomes.

[9] In genetic counselling geneticists typically resort to textbook illustrations, that is, to simple black-and-white representations. Science centres, museums, glass laboratories, and other popular science magazines, however, often draw on the power of colourful images. They strive to

Shortly after the bead of strings the geneticist begins talking about another visual representation, the image of the double helix. She points out to Ms M. that DNA is her genetic material. Further explanation is not considered necessary. "This is a... maybe it's an image you've seen before, this double helix structure", she remarks, and Ms M. answers: "Yes, yes, exactly." The geneticist here appeals to her client's existing knowledge of popular science. She builds the genetics lesson on everyday connotations. The double helix is a scientific model that has emigrated into everyday life as a symbol-laden image, thereby having enjoyed an especially giddy career. Ludwik Fleck described the development of a scientific fact in three stages: from an uncertain hypothesis in journal science to a certain scientific fact in textbook science to a seemingly universal fact in popular science. From the beginning the double helix, however, was not only a hypothesis but already a stable model, and then ultimately became elevated to a completely new kind of scientific image: to a cultural emblem or idol (Pörksen, 1997, pp. 105–35). In every imaginable colour and shape it appears on book covers and homepages, in company logos and advertisements.[10] It no longer has any scientific explanatory value; any

---

enthral their visitors with vividly colourful illustrations and computer animations. The less comprehensible and more abstract the scientific fact, the more jazzed up and artistic is its representation in popular science. The magazine *GEOkompakt* 'Humans and Their Genes', 2006, developed under the guidance of renowned scientists, is chock full of computer-generated infotainment on high-quality glossy pages. Artistic, three-dimensional images suggest that readers are being offered a glimpse inside a cell; what could be an outer space vision by Salvadore Dali represents how "DNA commands" migrate from the cell nucleus to the cytoplasm. Such an artistic world is intended first and foremost to evoke fascination and admiration: fascination for the mysteries of life, and admiration for the science that can unveil it.

[10] Nivea, for example, markets a "DNAge Anti-Age" facial cream for men with the claim that the ingredients protect the DNA against harmful environmental influences. A very abstract double helix is displayed on the product, see http://www.nivea.co.uk/products/mens-care/active-age [5.8.2014].

scientific value has long receded behind its social meaning. Charged with myriad connotations the double helix has become a symbol for the mystery of life and its scientific exploration: a spiralling stairway to progress. The geneticist draws on this symbolic meaning of the double helix when she appeals to something familiar in her talk. She builds her genetics lesson on popular associations such as the basic building blocks of life, fascinating knowledge, and technological malleability.

3.1.3.2. Reification through language

Not every image used by the geneticist to communicate genetic knowledge to her client is visualized. Some images remain on the linguistic level.[11] Yet their effect is similar to that of the visual representations lying on the table in front of the client: they turn scientific abstractions into seemingly concrete realities. While Ms M.'s geneticist draws on the bead of strings image to elucidate genes and chromosomes, one of her colleagues employs an entirely different analogy: he talks about chromosomes as a gene-containing "package". When showing a client a schematic illustration of a chromosome, he asks her to imagine this chromosome as the outside of an envelope containing genes:

> C: Chromosomes, they are the carriers that package genes; we may have around seventy thousand of them, and they are so

---

[11] It is common to analyse linguistic images in science and popular science as metaphors. Kovác and Frewer, in their study of genetic counselling, for example, talk about the objectifying power of metaphors (Kovác and Frewer, 2009, p. 208). I deliberately resist understanding these images or transferences as metaphors. Metaphors are "meaning vehicles"; they transfer meaning from one sphere to another and manifest themselves by means of analogy. Comparing genes with "administrative regulations" or the "genetic code" with a text, however, is something fundamentally different from talking about the "haven of marriage". "Genetic administrative regulations" links something that is fundamentally different and incommensurable. Analogies between the everyday world and the laboratory attribute a kind of meaning to scientific facts that they cannot have.

much smaller, the genes, that we cannot see them (-) under the microscope.

Like the comparison with beads on a string, this statement gives the impression that genes are little particles that can be isolated and counted. The human eye is just not powerful enough, suggests this counsellor, to perceive the reality of genes. Genes, he suggests, are just as tangible and real as a small grain of dust or sand. Only their minimal size makes it impossible to see them under the microscope. Such verbal expressions turn a scientific model or a heuristic aid into tangible reality. Genes appear in such representations with the same claim to reality as concrete and perceptible objects in our environment. Only the limitations of human perception, according to the geneticist, prevents us from seeing the reality of genes without science and technology.

Not all analogies and images are even recognizable as such. Transferences can also take on a life of their own and become consolidated into apparent facts at the level of jargon. An example of a congealed transference that frequently occurs in genetic counselling is "gene error" or "genetic defect".[12] While Ms M. received an in-depth lesson on the biochemical bases of gene mutations, most geneticists forego such details and allow the words to speak for themselves. Their explanations rest on everyday connotations of "error" and "defect". These connotations, however, are misleading: base sequences cannot be, like a human being or an organ, "healthy" and "defective" or even "normal" and "abnormal". An experienced observer can classify organisms, organs, or tissue based on their form and characteristics as "healthy" and "pathological".[13] Pathologists,

---

[12] The "genetic defect" is a very commonly used image in genetic counselling, see also Kovác and Frewer (2009, p. 213).
[13] Naturally, the distinction between "normal" and "pathological" is, like all scientific concepts and observations, historically variable. Ludwik Fleck wrote an insightful article on "Scientific Observation and Perception in General" where he argues that scientific observation requires specific training for acquiring the ability to perceive certain forms and

for example, microscopically examine histological slides and assess the morphology of cells. But at the level of base sequences there is no observable morphology. DNA-molecules cannot be "defective" or "pathological"; they do not possess any qualitative traits in this sense. Base sequences can only vary. A sequence variation must always be associated with a specific phenotype to be interpreted as a relevant "modification". Only when a DNA variation, which initially is nothing more than a meaningless difference, is correlated with an appearance regarded by medical specialists as pathological do geneticists interpret this difference as a mutation.

Talk about "normal" or "healthy" and "flawed" or "defective" genes is thus a misleading transference. In particular, the shorthand term "gene error" or "genetic defect" compares molecular variations with objects that can be flawed. Just like printing errors, construction errors, and material errors, apparently there are now gene errors. The gene is consolidated into a "something" and ends up on the same level of reality as a printed text or an unstable building. Even greater than with "gene error" is the reification drive with the collocation "genetic defect". An error can also be intangible, for example, an error in reasoning. A "defect", on the other hand, generally identifies material damage resulting in a loss of function. Talking about "genetic defects" suggests that a gene can be broken like a vacuum cleaner or a kitchen appliance.

Talk about "gene errors" and "genetic defects" encourages belief in a cellular microcosm wherein lurk invisible but consequential errors. Clients are asked to see themselves as carriers of such errors. As interviews show, such notions of carrying something defective inside one's self are very persistent. A few weeks after a genetic counselling session most people can no longer remember the details—not even the risk figures that are the main focus of the session. What persists is the fiction of

---

that this inevitably introduces a certain alogical factor into science, see Fleck (1986).

carrying something defective or bad inside oneself: "those bad things" or "the bad gene", as the interviewees formulate it (Lock, 2009, p. 76).

Everyday experience shows us that defects can typically be fixed. Thus talk about "genetic defects" suggests that there are also ways to repair things on the genetic level. In fact, for many years now geneticists have been promising a *technological fix*: gene therapy. "The most common form of gene therapy involves using DNA that encodes a functional, therapeutic gene to replace a mutated gene", proclaims, for instance, the internet lexicon Wikipedia (6 August 2014). After several patients died in the aftermath of such experiments, talk about gene therapy has died down. Yet geneticists still hold out the hope that genetic engineering will be able to repair defects. For instance, researchers from the Epigenome Network of Excellence want to "turn on", "turn off", and "modify" genes with epigenetic tools in the battle against cancer (Epigenome NoE, 2008). After years of media bombardment with such promises it is not surprising that one counselling client hopes that a defective gene can simply be removed. He is visiting a genetic counselling centre because several of his relatives have cancer. He himself had a brain tumour. The counsellor explains that sometimes gene errors can cause cancer. This error is what the client would like to have "removed":

> M: Can't the gene, uh, that's somehow responsible, or where it's not, can't it be found or and or can't it be removed or... [...]
> C: ((aloud)) No, no, it can be detected, even <u>analysed</u>, yes, that's possible
> [...]
> M: But it can't just be removed?
> C: But they can't do everything, but they can't treat it, manipulate it, that's not possible.
> M: Can they or can't they.
> C: No, no, I mean, at this level they can't.

The client must now live with the idea that he may be carrying a fatal defect about which he cannot do anything. Talk about a "genetic defect" turns him into a defective product that can be labelled as such but cannot be repaired. This imaginary defect and the alleged technological "not yet" creates a new form of suffering. From now on the man fears not only a recurrence of cancer but also suffers a neediness for which there is no technological solution. He is now considered defective and in need of repair; yet gene technology has not yet advanced to the point, at least according to the promise of technological progress, where he can be fixed.

3.1.4. Hidden causes

When I studied human genetics in the early 1990s, we were still being taught that genes cause diseases. Every textbook back then stated that an inherited disease such as cystic fibrosis[14] was triggered by a mutation in a corresponding DNA segment. The "gene" was considered to be the unambiguous "when" of a causal, linear "when-then" relationship.[15] The results of the human genome project, however, have radically challenged this idea of causative genes. Today, geneticists concede that the rela-

---

[14] Cystic fibrosis was once considered the epitome of a monogenetic hereditary disease. Genetic textbooks explained that in cases where both parents have a genetic mutation and both pass it on to their offspring, the child will have the disease. After the gene was localized in 1989 and examined, this model of a monocausal, linear, and definitive relation between DNA and disease broke down. Today, geneticists admit that it is not possible to predict the development of the disease on the basis of a genetic test. Even in the case of the so-called "severe" mutations delta F 508, the typical severe symptoms such as the thick, sticky mucus might not occur. "The variation on cystic fibrosis (CF) lung disease and development of CF related complications correlate poorly with the genotype of the CF transmembrane regulator (CFTR)" (Slieker et al., 2005, p. 7; see also Castellani et al., 2009).

[15] For the history of this trivializing understanding of causes in genetics, see Schwartz (2000). She shows that experiments did not investigate the relationship between gene and trait but rather between gene mutations and trait changes. If a certain mutation changes eye colour, this does not mean that the sole cause of eye colour has been discovered. In theory formation, however, the changing factor becomes the cause, the "gene for". "Theory and research were not guided by the same premises" (Schwartz, 2000, p. 31).

tionship between DNA and disease is by no means as monocausal, linear, and unambiguous as they had believed for decades. Normally a disease cannot be diagnosed based on a DNA sequence; no one can predict whether an unborn child or an infant with a so-called gene for cystic fibrosis will ever develop the typical symptoms. The same is true of other "monogenetic" classified diseases: some with the corresponding "gene defect" will become very ill while others will remain healthy their entire lives.[16]

Whether the "drinker gene", "smoker gene", "breast cancer gene", or "gay gene" — almost every "gene for" that has been loudly proclaimed in recent years actually announces nothing more than the fact that a geneticist has perceived an often dubious statistical correlation between genotype and phenotype. What such a correlation reveals is the following: certain genetic markers and certain human traits, whether they are diseases or undesirable habits, appear together so frequently in various random studies that they are statistically dependent. Nothing more. If the sample or other study parameters change or if a different statistical procedure is used, then often the statistical relationship disappears. In any event, a statistical correlation says nothing about cause and effect in the conventional sense. For example, there is a strong correlation between the decline in the number of stork nests and the birthrate in East Prussia. If the belief in storks delivering babies was as prevalent as the belief in genes, then this correlation would be regarded as a causal relationship. Further questions would not even be asked. The basis for this correlation, namely the rise in industrialization, would never become an issue. But this cause, so plausible to us today, cannot be proven by statistics. If "cause" appears in statistics, the word does not identify a causal

---

[16] Thus, for example, the geneticists Ruth Hubbard and Richard Lewontin talk about two sisters to whom the same "gene for" a hereditary eye disease (Retinis pigmentosa) was attested: one sister is blind, the other drives a truck at night, see Hubbard and Lewontin (1996).

relationship but rather the consistency of a probability, as the French philosopher François Ewald summarized the concept of cause in statistics (Ewald, 1993, p. 183).

In genetic counselling, however, such a correlation between DNA and its manifestation becomes an alleged causal relationship. As soon as the genetic counsellors convey their expertise in universally comprehensible, everyday language, DNA becomes the cause. This implicit causality is already embodied in the words "genetic defects" and "gene errors". Without even being questioned "genetic defect" makes DNA the cause of the disease—like a cylinder defect is the cause of an engine breakdown. Talk of "gene errors" and "gene defects" thus allows geneticists to evade the question of the complex and largely ambiguous relationship between genotype and phenotype.

But geneticists are not only speaking implicitly but also quite explicitly about genes as a cause. Ms M.'s geneticist very clearly equates the "gene mutation" that Ms M. may carry with a cause:

> C: [They] would [...] go through these genes and look through them base for base to see if this, if everything is normal or if there is a mutation on some position that could cause a disease.

An internal cause that may cause cancer one day is obviously not something Ms M. can imagine. For her there is no difference between the disease and the so-called gene error. She believes that the gene mutation already is the feared cancer disease. While the geneticist is lecturing about DNA mutations, she abruptly interrupts her and asks:

> W: And this will wake up someday? Suddenly, or what?

Here Ms M. equates the mutation with colon cancer. She declares the biochemical "mutation" of DNA as the beginning of disease—as something that is slumbering latently and then "wakes up" or actualizes. The gene mutates and sets the disease in motion, she believes. The counsellor corrects her client. She clarifies that she did not mean the mutation was a health impairment that started at a specific time but that it is a basic error, something that has always been there:

C: Noo, that is, it's like this, that one has... that it is a so-called germline mutation, that means it has been there from the beginning.

The geneticist is talking about an apparent cause of disease that may be a constitutive part of Ms M. As she further explains, this cause of disease would be everywhere and would always have been there:

> C: [...] is also detected in every cell of the body. So that means if there is a specific mutation, for example, in a gene associated with colon cancer, then you would have one copy of this mutation, in other words, on one of both chromosome segments, associated, from the beginning and in every cell. So at the very least one could, so to speak, tear out a hair, DNA can be extracted from a hair root and then they can look to see whether, um, whether a mutation is there or not, or draw blood or something.
> W: Yes.

Unlike an abscessed tooth, a crippled leg, or an irritated stomach, a genetic flaw does not have a beginning in time nor can it be localized in the body. If this mutation is detected in Ms M., then a constitutional defect will be attested to her from which she will not be able to distance herself. An error that cannot be spatially or temporally isolated, that she cannot perceive or pin down. An error that cannot be treated or alleviated, neutralized, or removed. An error that has existed before her, namely in the genes of her ancestors, and which is more or less nowhere and yet everywhere. A defect that would exist in the foundation of her being.

3.1.5. Meaningful information

The existential meaning of "genes" and "gene errors" is reinforced by the notion that something is being stored there — "genetic information". Gene mutations, clients are taught, are mutations in the genetic code, that is, in the person's blueprint. A young pregnant woman, for example, is taught that genetic information contains information about the entire person:

"Every cell has all the information about what constitutes the person... This means that when I remove <u>one</u> cell from someone's body, I have representative genetic information about the entire person" (Samerski, 2002, p. 157). Thus genes have intangible content; they contain information, instructions, or "administrative directives" (Samerski, 2002, p. 156). The same counsellor who spoke about chromosomes as "packaging" compares genes, for instance, as having the content of an audiotape:[17]

> C: Chromosomes [...] can be compared with an audiotape that can't be read without a medium. When it's the genes in chromosomes, we also need a medium. A different kind.

Genes here are defined by and large with what they store, that is, as something that can be "read". This "reading" is merely a question of having the right technical medium, according to the geneticist. Thus the client is to understand that she has a small information storage system inside her, whose content and meaning can be accessed with the right device.

For approximately half its lifetime the gene did not contain any "information". Until the mid-twentieth century geneticists assumed that the gene was determined by chemical or physical reactions. "Genes did not transfer information before the 1950s; they only possessed biochemical specificities", as Lily Kay explains (Kay, 2000, p. 18). This changed when cybernetics[18] and

---

[17] Comparison with an audiotape is also popular elsewhere; see, among others, Kovác and Frewer (2009, p. 214).
[18] Cybernetics is a discipline, or maybe even a worldview, that arose from the US think tanks and laboratories of industrial and military research in the 1940s. Mathematician Norbert Wiener (1894–1964) worked on the improvement of the anti-aircraft defence when he modelled the first man-machine-system and thereby laid the foundation of this novel science of control. Wiener celebrated cybernetics as a new *Weltanschauung*; he transferred his ballistic feedback model to all "purposeful behaviour" and popularized key concepts such as "communication", "control", "behaviour", "input", and "output" as a new foundation for the control of man and society (Wiener, 1950). Even though the cybernetics enthusiasm of the 70s has declined, cybernetic

information technologies began their triumphal procession. Molecular biologists were fascinated by the technological and epistemic innovations in the information sciences and transferred their terminologies to their own field. Around the same time, as the first commercial computers were leaving the assembly line, geneticists began understanding the gene as a program containing instructions. Soon thereafter the organism became a cybernetic system, genetic material became an information storage system, a biochemical base sequence became a "text" or "code", and genetic make-up became a genetic program. "The gestalt switch to information thinking in biology, with all its paradoxes and aporias, was even more fundamental than the subsequent (1953) paradigm shift from protein to DNA" (Kay, 2000, p. xv).

However, concepts from information theory that are precisely defined in mathematical theories, in algorithms, cannot be so simply transferred to the study of life. With their emigration to biology they underwent a radical change: they lost their technical denotation and became charged with everyday and popular scientific connotations. In information theory "information" is strictly defined: it identifies a measure for the selection or reduction of equiprobable options and is thus purely quantitative. "Information" is something purely formal; there is no meaning in information theory. "The word 'information' in this theory is used in a special sense that must not be confused with its ordinary usage. In particular, information must not be confused with meaning." Thus two messages—warns communication scientist Weaver—one full of meaning and the other utter nonsense, such as a Shakespeare sonnet and a random collection of letters, can be fully equivalent from the standpoint of information (Shannon/Weaver, 1949, cited in Kay, 2000, p. 99).

---

precepts and concepts have transformed knowledge and technologies from sociology to engineering to the life sciences.

Even tenacious attempts to transfer and quantify information in this technical sense to biology fail miserably.[19] "Genetic information" has nonetheless advanced to become the leading concept of genetics, even though it does not have much in common with the information of information theory. It receives its plausibility and meaning primarily from everyday connotations and means something like the objectifiable content of a text or instruction. All it has taken from information science is the appearance of objectivity and scientific rigour. Meanwhile the concept has taken on a life of its own and the memory of the transfer has faded, with the result that "genetic information" appears as a scientific fact, even as something natural. Therefore genetic counsellors do not find it necessary to explain it: they speak matter-of-factly about "genetic information" in the genes or in DNA — about the information which "constitutes the entire person".

Genes, as genetic education implies, are the starting point, the origin, and the cause of our own existence. They store basic information and exist everywhere. Unlike with dementia, liver cirrhosis, and muscle weakness, genes and gene mutations are not something someone has; rather they are what someone is. The attribution of "something genetic" is thus much more powerful than a conventional medical diagnosis. Genetic education implies that geneticists shed light on the invisible truth about one's self. They reveal a hitherto undetected "identity":

---

[19] The biologist and physician Henry Quastler tried in vain, "in a technically proper form" (Kay, 2000, p. 114), to found an information-based biology and, although his "discursive framework survived and flourished", he himself has fallen into oblivion. See Kay (2000, pp. 115–27). Just as unsuccessful as Quastler, incidentally, was the "decoding" of the genetic code: for years during the Cold War geneticists, mathematicians, and cryptographers were pulling their hair out trying to break the code. Only hard work by biochemists, namely the tedious experimental arrangement of amino acids into the base triplets of DNA, finally produced the famous "codons sun". "From linguistic and cryptoanalytic standpoints, the genetic code is not a code; it is, rather, a powerful metaphor for the correlations between nucleic and amino acids" (Kay, 2000, p. 11).

"Genetic disease differs in as much as it promises to reveal who the individual always has been, not a new addition but a revelation about an underlying identity that had been concealed" (Armstrong, Michie and Marteau, 1998, p. 1658). Ms M. receives such a genetic identity. She learns that since birth she might have been carrying a consequential cause of disease in her, an error in her genetic program. And with this revelation she is no longer who she was: she has become a gene carrier, a phenotype of a defective genotype. And not only has her understanding of herself changed but also the cancer disease she fears. Something happens to those who unexpectedly develop colon cancer. The disease is a stroke of fate that unexpectedly changes one's life. However, if Ms M., as a gene carrier, develops colon cancer it is no longer something that happens to her but something that appears to be stored in her. She has always carried the disease in her as a defective blueprint or information unit; the disease would be the expression of her defective genotype. Thus the gene reframes what could happen to her into something she is. No wonder that a woman, after having been informed about genetic mutations correlating with breast and colon cancer, in an interview stated in retrospect: "I was thinking, what other genes are also defective?... I wanted a new identity, I didn't want to be the person I was anymore" (Porz, 2009, p. 56). She lost her self-confidence and felt socially alienated.

3.1.6. Internal agents

So far the geneticist has spoken about genetics in general. When she turns to genetic testing, she begins conjecturing about her client's genetic make-up. She speculates about the specific gene errors and disease genes that Ms M. may have. In this section of the gene lesson more phrases occur in which the gene does something. The gene here is no longer an invisible thing, an information storage system, a readable directive, a constitutive defect, or a passive cause. Instead it has become an agent. The geneticist formulates numerous sentences in which the gene is the subject of a verb indicating a causative activity. Ms M.

learns, for instance, that she may have a gene mutation that could "result in" a disease:

> C: And this mutation can specifically result in (-) specific diseases, in other words, it can trigger them.

The (gene) mutations "result in" diseases or "trigger" them, according to the geneticist. Such formulations are very prevalent in technical language as well as in popular science presentations and look like objective descriptions. Yet, like the phrase "gene defect", they convey meanings from everyday language to the abstract sphere of markers, statistical associations, and probabilities. The "gene defect" has already turned a statistical correlation into a "cause". Genes that "trigger", however, suggest not only an if-then relationship but active causation. In the process, the gene takes on animist traits: it becomes the mover that initiates acts on its own. Genetic counselling transforms a genetic marker, that is, a statistic feature, into an object that can actively cause. An educational lesson turns a mathematical relationship that a client would find unintelligible and meaningless into a "gene" that does something.

At other times in the genetic counselling session the geneticist very explicitly ascribes to genes the power to actively make something happen. Genes appear in the role of doers or agents. Gene mutations "cause" the disease, as the counsellor declares in the following sentence:

> C: But there are mu... um, gene mutations, which we know, uh, that mainly cause only breast cancer.

Other geneticists talk about genes "causing", "triggering", "regulating", and "generating". Genes appear as the agents that do things to people. They become perpetrators, and people become the suffering victims. Genes receive the status of subjects, and people become their objects.

Conceiving genes as subjects is carried to the extreme in popular science, whose goal is to fascinate its readers with lively and vivid language. The *GEOkompact-Heft* "Humans and Their Genes" (2006), for example, written with the assistance of numerous international geneticists, teems with sentences in which genes, in myriad ways, are active: genes "shape" behaviour, they "regulate" development, they "determine" our inheritance, they "see to" the distribution of hereditary information, and "protect" us from diseases. Often people are depicted doing something but the real agents are shown to be the genes: a businessman pointing a pistol at a streetboy kneeling on the floor. Genes determine individual levels of aggressiveness, the author comments (Engeln, 2006, pp. 10–11). On another page half-naked women and men greet one another on the beach. Even in matters of love, we read, genes direct our behaviour (Engeln, 2006, pp. 12–13). A person, so it seems, is only the external façade of invisible powers. She only carries out what genes dictate to her: the theme of the nine articles on genetic topics such as "Men and Women", "Altruism and Aging", is therefore "how genes control us" (GEOkompakt, 2006, pp. 52-3).

3.1.7. Genes as an "illusion"

The goal of the lessons on "genes" is to teach Ms M. about herself. To do so, the geneticist first points out to her client that, like all people, she is a gene carrier. She explains genes, DNA, chromosomes, genetic information, and mutations, and asks her client to see herself as the product of these invisible gene worlds. Subsequently, she ascribes to her specific genes and gene mutations, namely, such that "trigger" or "cause" colon cancer. The whole time the geneticist talks about genes in a way that turns the "gene" into a tangible and forceful reality. The linguist Elizabeth Shea concludes that, from the beginning, since the term was coined by Johannsen in 1909, it has been conceived as a type of metonym: as a word that identifies an abstract and complex phenomenon as if it were a physical thing (Shea, 2001). Philosophy calls the process of objectifying and personalizing a thought construct in this manner hypostatization. Immanuel Kant says that we "hypostatize that which exists merely in thought and thus assume it to be a real object outside of the thinking subject" and calls this a "mere mirage" (Kant, 1999, p.

434).[20] In genetic counselling the gene is such a mirage, a hypostasis.

If one listens to the genetic experts, genes are not only tangible and small, they also have impressive traits and prowess. Genes, as a diachronic receptacle for the popularization of consecutive stages of science, are endowed with myriad qualities: they are "small" and "everywhere", divide and recombine, "mutate", are "lined up", "packaged", travel through generations, are "defective", are "copied", "store information", "contain readable material", are "active", "trigger diseases", "control", "cause", and are "responsible". There is likely no other biological construct that can be linked in everyday language with so many different adjectives and verbs. Inevitably, the gene has been inflated into a fateful and omnipotent agent—into the "quasi-mythical entity" (Keller, 2000, p. 145) that has determined thinking about life in the twentieth century.

It would be misleading to discount this hypostatized and mythical gene as the popular science distortion of a real, objective gene in the laboratory. For one, because this real, objective gene in science does not even exist. "Gene" has a precise meaning only within very specific experimental practices, in other words, when researchers using the same methods are working on the same problem.[21] Thus, in genetic research

---

[20] In German: "bloße(s) Blendwerk" (Kant, 1781, A 384).
[21] "Moreover, it is from the specificity of the experimental context in which they are invoked that technical terms acquire the precision they need. Terms like *gene* may be subject to a variety of different meanings; but locally, misunderstanding is avoided by the availability of distinct markers directly and unambiguously tied to specific experimental practices. Within that practice, the marker has a clear and unambiguous reference. Change the practice, and different markers will need to be employed. And inevitably, these different markers will pick out somewhat different physical entities. Nevertheless, as long as one stays within the context of a given and clearly understood set of experimental conventions, the term *gene* can still safely serve as an operational shorthand indicating (or pointing to) the marker with immediate experimental significance" (Keller, 2000, p. 140).

laboratories there are many different genes that cannot be reduced to a common denominator. Science philosopher Philip Kitcher sums up the status of the term gene in genetics: "A gene is anything a competent biologist chooses to call a gene" (Kitcher, 1992, p. 131). On the other hand, as Ludwik Fleck understood, popular scientific belief in the reality and significance of his research findings is the motivating force for every scientist. Popular science is therefore always a constitutive part of scientific concept formation (Fleck, 1981). Thus it is no surprise that geneticists, despite inconsistent and contradictory research findings, have proven to be particularly strong believers in genes. The gene has advanced in the twentieth century to become a powerful thought constraint, not only in the public realm but also in science. Today the genetics project still lives on in the "genes in the heads" (Duden, 2002a): from the belief that there are genes that determine the phenotype and regulate the organism, that genetics is encoding the mystery of life and ultimately will contribute toward creating a better world. These convictions have spurred genetic research and dictated research questions, methods, and research findings.[22] The hypostatized, omnipotent, subjective gene is therefore not only the result of popular science communications, but is also the ideological complement of the scientific gene industry. And this ideological complement, the "gene in the head" of geneticists, appears in genetic counselling sessions. When geneticists leave the domain of their specialized language and bring their

---

[22] "Science in Context" studies and studies in the history of science have demonstrated in detail that the natural sciences are not free from social and cultural "contamination", but rather are highly dependent on their social and cultural context. Not only does Pearson's correlation coefficient have its roots in eugenics or can the formulation of Bohr's complementarity principle be attributed to the *fin-de-siècle* philosophy of life (Daston, 1998, p. 22), but also the criteria for scientific objectivity has such social historical origins (Daston, 1998). See also Shapin (1994). On the history of objectivity and the endeavour to understand science from a cultural historical perspective, see, among others, Daston (1998; 2000; 2001), Daston and Galison (2007).

expertise into everyday life, they constantly talk about this ideological gene. As soon as geneticists attempt to depict their expertise as relevant to everyday life, they inevitably fall back on their own "genes in their heads": they ascribe to their clients those genes which correspond to popular scientific notions, their genetically based worldview.

3.2. Second Transformation of the Person: Clients as Risk Carriers

After about twenty minutes Ms M. has the gene lesson behind her. She has learned to see herself as a carrier of significant, influential genes that can mutate. She has learned that complex tests can ascertain such gene mutations. However, what all this means for Ms M. personally has not been addressed by the geneticist. So far she has merely clarified that everybody is made of DNA, chromosomes, genes, and genetic information, and that with Ms M. a consequential error may have slipped in. But she has not yet explained what this would mean for Ms M.'s life and feared colon cancer. Now the geneticist will change this: she announces that their conversation will take a turn. She apologizes for the long lectures about DNA, chromosomes, and mutations, and promises to be "concrete" again—that is, she acts as if there is now something tangible and significant to say about Ms M. Then she reveals to her client that she is a "high-risk person":

> C: To stop bombarding you with (laughs) information, I think, we will now be concrete again about what... what we can now tell you.
> W: Yes
> C: So! Based on the criteria we can tell you that you belong to the category of high-risk persons.
> W: ((looks startled)) Hmmm. ((laughing a little, loud)). That sounds bad, high risk!
> C: Yes, yes, or, in other words, an elevated risk.
> W: Yes, yes.
> C: For hereditary colon cancer, um, and theoretically we can offer you this molecular genetic testing.

W: Hm=hm.

After the crash course in genetics, as the geneticist's transition promises, she now wants to talk about what geneticists — here she uses a professional "we" — can offer her. And what does she do? She ascribes a risk to her client. Or, more precisely, she classifies Ms M. as a "high-risk person". And although Ms M. already believes that she will probably develop colon cancer, as she says elsewhere,[23] and has already had a polyp, she is stunned by this statement: "That sounds bad, high risk", she responds with a shocked look on her face.

3.2.1. A grave misunderstanding: Risk as diagnosis

Ms M. is obviously shocked because she believes that with the risk classification the geneticist is saying something concrete about herself. After all, the expert heralded the risk attestation as a concrete statement. Ms M. does not realize that she has only been statistically pigeonholed. She thinks that the expert has revealed something significant about her health status. She believes that her statements have rhyme and reason. She does not realize that the health professional has merely ascribed to her the probabilistic traits of a fictive cohort of patients.

Statistical probabilities have a very precise, but limited, validity. They say something about what happens in a long series of similar experiments or events — but they do not predict the outcome in a single experiment. By definition probabilities quantify frequencies in populations but make no predictions about individual cases. The example of gambling, the cradle of probability theory, illustrates the chasm between the individual

---

[23] W: I didn't know how high the risk was, what percent, and so I thought: okay, it's hereditary with us, it occurs so often, somehow it became clear to me: we have a problem with colon cancer. Now I could at least be told: your risk is not that high. For me, things can only get better because everyone else has already had it and uh=
C: = You simply assume that you'll get it=
W: = Yes, yes, exactly.

case and series: only when I roll the dice very often, several hundred times, do statistical regularities appear. Only with many repetitions of a so-called random experiment, which quantifies the so-called law of the largest numbers, do the empirical frequencies come close to the calculated probabilities. When throwing dice each number of the die has a probability of 1:6. Which number of the die I will get when throwing the die once, five times, or ten times, however, cannot be predicted. Even with fifty or a hundred rolls of the dice the number of the die is not equally distributed; the one and the three may appear disproportionately often while the two appears less frequently. Only after many rolls of the dice will all six numbers of the die gradually appear with equal frequency, thus approximating their calculated probability of 1:6. However, if I only roll once, lady luck rules—otherwise why would anyone hope for great fortune?

The early statisticians in the nineteenth century were aware of this gulf between the calculated and the concrete, between the regularity of the measure and the individual case. Although scholars such as Francis Galton or Adolphe Quételet (1796–1874) enthusiastically measured, counted, and calculated everything—from horse steps, the effect of prayers, to the breadth of the male chest—they did not perceive their fellow humans as risk profiles. The Belgian mathematician Quételet founded social physics, which declared social phenomena as statistical regularities, and invented the "average man" ("l'homme moyen"), a statistically constructed norm by which the individual would henceforth be measured (Ewald, 1993). Nevertheless, Quételet expressly warns against drawing conclusions about individuals based on statistical laws: he clearly stated that these laws, in accordance with the manner of their determination, no longer have anything individual about them. Any application to an individual person would essentially be false; it would be like using a mortality table to determine when a certain person will die (cited in Ewald, 1993, p. 196).[24]

In genetic counselling these two levels, the regularity of the masses on the one hand and the individual case on the other hand, are systematically blended. The geneticist talks about risks and probabilities as if they were something concrete, generally comprehensible, and personal. Statistical calculations and concrete, tangible threats are thrown into one pot. The "elevated risk", which she attests to her client, suggests it would quantify Ms M.'s disease susceptibility. With her statement "You belong to the category of high-risk persons" it seems as if the counsellor is making an alarming diagnosis.[25]

But the attested risk does not say anything about Ms M. and her future. For Ms M. the disease probability is about as irrelevant as the probability of rain for today. According to the newspaper, there is an 80 percent chance of rain, but now in the evening there are still no clouds in the sky. It would be misleading to make a conclusion about the actual weather based on

---

[24] Peter Bernstein (1996) has written a very detailed and informative history of risk and probability thinking. Lorraine Daston (1988) examines seventeenth- and eighteenth-century theory of probability whose goal was to describe rational thinking—"good sense reduced to calculus." Ian Hacking (1990) has since elaborated a pertinent history of probability theory and statistics in which he explores the social and cultural effects of this new thought form. The emergence of probability thinking and its effects in various scientific and social areas has been investigated by Krüger, Daston and Heidelberger (1987); Krüger, Gigerenzer and Morgan (1987). A concise summary of their findings has been published in Gigerenzer et al. (1989).

[25] Whenever risk emigrates from statistics to the doctor's office or to the sickbed, it undergoes a radical transformation: physicians (and patients) interpret risk factors as "objective clinical signs of disease", as Gifford (1986, p. 222) writes. Such "clinical risks" are not, strictly speaking, real risks, as Lorna Weir points out, because the future of a specific patient cannot be calculated at all; they are based on an amalgam of irreconcilable forms of thinking, namely, the ascertainment of risk factors on the one hand and the diagnosis of normality and abnormality on the other hand: "Clinical risk comprises an unstable amalgam of incompatible forms of reasoning" (Weir, 2006, p. 19).

rain probability: it does not give us any information about whether rain will fall or not, for how long and when, nor whether it will drizzle or come down in a downpour. Instead, what the 80 percent tells us is the following: yesterday the meteorologist constructed a fictive "today" in which it rains 80 times out of 100. If there were 100 days after yesterday, it would rain—sometime and somehow—on 80 of those days. In reality there is only one "today". And for this one "today" 80 percent is irrelevant.[26]

3.2.2. The client as a statistical construct

Similar to how a meteorologist constructs a fictive "today", the geneticist constructs a fictive Ms M. Out of the concrete person nervously sitting across from her, she makes a risk profile—a statistical double. For this purpose she has, at the beginning of the session, taken down a slew of data or traits; based on these traits she then assigns Ms M. to a statistical collective and attests to her the probabilistic characteristics of this collective as "her risk". Hence, in order to attest a risk to her client, she must first radically reframe her: she must transfer her from the concrete, tangible reality of life into the realm of probabilities. A flesh and blood Ms M. is turned into an abstraction, a faceless risk profile.

Typically, a genetic counselling session begins with a detailed enquiry of the client. These questions serve to capture data necessary for the construction of a risk profile. This is also the case with Ms M.—the geneticist ticks off a whole checklist of questions to capture specific criteria such as the occurrence of colon cancer in her family and her own previous history with

---

[26] Cognitive psychologist Gerd Gigerenzer has devoted his recent research to the communication and understanding of risks. In his book *Reckoning with Risk: Learning to Live with Uncertainty* (Gigerenzer, 2003) he clarifies common misunderstandings that are prevalent among laypeople and experts with regard to probabilities and risks. He emphatically points out the precise but limited informative value of test results and statistics while making a case for critically evaluated probability calculations as the acme of enlightened thinking.

polyps. She asks for the age of her client, her previous health status, as well as any medical findings and diagnoses. Subsequently she extends the questions to all of Ms M.'s relatives. During her registration Ms M. already filled out information about her family, and it now lies in front of the counsellor. They go over this information again in detail. The geneticist asks about diseases and possible causes of death for parents, grandparents, aunts, uncles, nieces and nephews, and their children. Men are marked on the paper as squares and women as circles, so that the geneticist draws a whole cluster of linked geometric figures. This family tree will function as the basis for Ms M.'s risk classification. The geneticist is searching for markers or abnormalities that are relevant to hereditary colon cancer, that is, for certain cancer diseases and colon polyps. This is her grid through which she looks at Ms M. and her relatives. She is not interested in anything else because it is irrelevant to the construction of a risk profile.

Ms M. knows nothing about this grid and eagerly chats away. She talks about what seems important to her with regard to the disease of her relatives and her own insides, and relates this in the form of anecdotes and stories. She reports about her "irritated intestines", which react sensitively to stress and cold beverages, and about the fact that a physician told her she had a particularly long colon. But the geneticist can do nothing with all these stories through which Ms M. imparts meaning to her life and her experiences. Personal stories are irrelevant in the grid of "genes" and "risks"; only standardized, objectifiable information is important here. The geneticist wants to capture very specific data and therefore asks about disease diagnoses, age of onset, degree of kinship. From her client's stories she takes out what she needs to construct a risk profile. Scrutinizing the family tree she peers at the cluster of circles and squares and ascertains:

> C: Then, um, yes, so when I look at this, then it is naturally, as it looks, there appears to be a high incidence of colon cancer on your mother's side of the family.

The geneticist takes her client's account of disease and death and translates it into a marker: into the statistical feature "high incidence of colon cancer". The tragic stories about Ms M.'s relatives who died of colon cancer at an early age are seen as a mere classification criterion. This abstraction of concrete life circumstances and experiences is not insensitive but professional. It would be misleading to accuse the geneticist of ignorance. In order to assign a risk profile to Ms M. she has no other choice but to not take anything concrete into account. She must convert experiences and stories into data and criteria through which she can place her client in a population. The probability theoretical characteristics of this population are then assigned to her client as risks. She therefore attests to her a disease incidence that actually refers to a population.

This generation of a risk profile is the geneticist's main task. Not Ms M. personally but her statistical double is the pivot of the counselling session: from the risk profile the geneticist derives the future she predicts for Ms M., how menacing this future looks, which monitoring measures she recommends, and whether she will offer her genetic testing. However, the expert bases not only her own thinking on statistical abstractions, she also asks her client to do the same: Ms M. is to learn to see herself as a risk profile — and to make her decision from this new perspective.

What this new perspective on the self is and the mode of thinking it presupposes becomes clear when the geneticist explains how she classified Ms M. as a "high-risk person". There are two international sets of criteria, the so-called Amsterdam criteria and the Bethesda criteria, she explains. People, and thus potential testing candidates, are classified based on these criteria.[27] Ms M. does not meet the Amsterdam

---

[27] In the counselling sessions about colon cancer and breast cancer, family history is a basis for the risk classification of the client. As medical sociologist Ellen Kuhlmann points out in the case of breast cancer, however, "there is considerable room for interpretation as to how many

criteria, but does meet the less strict Bethesda criteria of "when two first-degree relatives had cancer... and the age of onset for one of them is under 45 years of age or the cancer was colorectal adenoma [colon polyp, S.S.]... before 40 years of age", says the counsellor. "So these criteria have been met because several relatives have been, so to speak, afflicted." But Ms M. herself also manifests one trait with which she fulfils a Bethesda criterion: a polyp before she was 45 years old.

> C: So in principle you yourself with your disease at 40, so to speak, meet the criteria.

The two international, standardized sets of criteria are instruments for constructing a population with an elevated probability of a familial disposition for colon cancer. The individual criteria are not symptoms suggestive of an underlying disease but risk factors for generating risk populations. Ms M.'s polyp, which definitely could be considered a symptom, counts in this context only as a risk factor. Therefore the criteria serve as a sieve to filter out potential test candidates from the general population. Similar to a dragnet investigation, a trait grid is used to generate a target group for which closer examination, here genetic testing, is worthwhile.

> C: Um (-) and based on this we can say that the family anamnesis already (--) suggests that something hereditary could exist.

The geneticist emphasizes "could". Ms M. has now been placed in a class in which "the probability exists that you have this hereditary form", as the counsellor states elsewhere. Based on her family tree and her previous history Ms M. has been categorized as someone with an elevated probability for a genetic disposition and a genetic risk—strictly speaking, that is, an elevated probability for an elevated probability. Based on a

---

disease cases and which degree of kinship substantiates the classification of a woman into a high risk group" (Kuhlmann, 2002, p. 86).

trait grid, an international set of criteria, she is classified as "suspect", as a "risk person". She thus belongs to the target group of potential genetic testing candidates. And for this reason she is being asked to make an "informed decision" about whether she will undergo the testing or not (see section 3.3).

The geneticist has verified whether she could ascertain certain traits with Ms M. and then, based on these traits, generates a risk profile, a statistical abstraction. She assigned this risk profile to her client, as something allegedly personal, as an ostensible barometer to measure her prospects, whether she is facing the same tragic fate as her uncle or her aunt. However, the risk profile has nothing to do with the person sitting there at the table, healthy and with an enquiring gaze. Everything constituting Ms M., everything concrete, unique, and physical, has vanished from the risk profile. What remains is only what is statistically ascertainable: a set of variables. Abstraction is not ignorance, but the precondition for statistics: "In statistical affairs... the first care before all else is to lose sight of the man taken in isolation in order to consider him only as a fraction of the species. It is necessary to strip him of his individuality to arrive at the elimination of all accidental effects that individuality can introduce into the question" (Poisson *et al.*, 1835, cited in Hacking, 1990, p. 81). Statistics, in principle, does not deal with concrete persons but only with faceless cases. The geneticist has turned Ms M. into such a faceless case, a "fraction" of the human species. She has epistemically transformed her client. This transformation makes it possible to equate Ms M. with other statistical doubles and throw her in a pot and make a statistical calculation.[28] And after this calculation the geneticist is finally able to ascribe to her the disease probabilities of this statistical "pot" or population as her alleged personal risk — and to suggest that by doing so she has made a significant statement about her person.

---

[28] The geneticist here does not carry out the calculation herself; but she is the basis for being able to calculate a probability.

3.2.3. The pathogenic effects of physician-attested risks

The idea that people can have risks is a relatively new phenomenon. The generation of my grandparents still feared concrete dangers but not quantified risks. The risk that can be anticipated, calculated, and insured has long been reserved for merchants and insurance. The German "*Risiko*" (first spelled *risico*) appeared in the sixteenth century as a commercial term, as did the English "risk" in the seventeenth century. The word emigrated from business and insurance to German colloquial language only at the beginning of the twentieth century, where it gained acceptance as a synonym for "danger" and "daring". In 1934 there was talk about the risks inherent in street traffic, and in the 1960s health apostles promoted filter cigarettes as "risk-free tobacco products".

Nowadays, in contrast, the epistemic transformation of persons into risk profiles is routine. Doctors' offices are filled with people robbed of their sense of well-being not by an adversity but by a risk prediction. Whether pandemics, early aging, an exceptional child, or lumps in a breast—in the "risk society"[29] everything that may happen is anticipated as a risk. However, "risk" does not identify a concrete reality but only a specific form of objectifying potential events. Risks in themselves do not exist. Conversely, this means that everything can be made into a risk: "Nothing is a risk in itself. There is no risk in reality. But, on the other hand, anything *can* be a risk" (Ewald, 1991, p. 199).

---

[29] The term "risk society" goes back to Ulrich Beck and his book of the same title published in German in 1986, see Beck (1992). The strengths and weaknesses of his analysis and his concept of risk have meanwhile been much discussed; see, among others, Wynne (1996). The term "risk society" is used here not only in the Beckian sense but also to poignantly identify a society in which administration and policy have made their main task the ascertainment, calculation, reduction, and distribution of risks—whether in fighting crime, in economic policies, in the health care system, or in social policies.

For the most part it is left in the shadows what this form of objectification of possible events as "risks" theoretically presupposes, namely the generation of faceless cases. Therefore the epistemic transformation of clients into risk profiles almost inevitably leads to epistemic confusion: to the confusion of the "I" or "you" in a colloquial statement with the statistical construct upon which the geneticist is basing her statements. The geneticist is personally talking to her client; she addresses her explanation of "genes" and "risks" directly to the woman sitting across from her, to a "you". But the addressee and referent of her statements are miles apart: what she is saying does not refer to Ms M. but to a statistical construct. Probabilities by definition do not refer to a concrete person but to a constructed "case"; never to an "I" or "you" in a colloquial statement, always only to a "case" from a statistical population.

The epistemic confusion between a statistical construction and a colloquial "you" is not only misleading but also pathogenic. When Ms M. sought out genetic counselling she was afraid that she may develop colon cancer. Nevertheless she is still shocked when the counsellor categorizes her as a "high-risk person". "That sounds bad", she responds. Obviously the risk attestation has conveyed a different message to her than what she already suspects: she may develop colon cancer. Since the counsellor has portrayed her risk as a concrete statement, even as a type of diagnosis, it must seem as if Ms M.'s future is no longer open. The risk attestation creates the belief that she will soon have the disease. Risk has turned the mere possibility into latency. The "either-or" – it may happen or not – has been turned into a "not yet". The anticipated future, predicted by analogy to a game of chance, is turned into a concealed present. Ms M. is thereby transformed into a new type of patient: she suffers from a prediction. She can no longer feel healthy but lives instead with a peculiar new condition: she is not yet sick.[30]

---

[30] The literature actually calls at-risk healthy individuals "the pre-symptomatic ill" or the "healthy ill".

This new condition of the constructed "not yet", in which people are placed by risk attestations, is pathogenic. The level of quantified risk plays no role here; though it is important for geneticists and statisticians, it is not for the person concerned. As interviews reveal, soon after the counselling session most people forget the exact figures, that is, in so far as they were able to remember them at all.[31] Much more important for them is the question of whether they are "normal" or not.[32] Even a low risk can make people sick because it conjures up a menace and questions their health status: "If people have an all-or-nothing perspective of harm or contagion, this concept [*of risk*, S.S.] will seem unfamiliar. The idea of safety is zero risk. Anything else is seen as dangerous" (Adelswärd and Sachs, 1996, p. 1181). Since nobody in their right mind can think about their own life in terms of probabilities, a risk is often transformed into a certainty. Those who have been attested a risk, for instance, assume that they will meet their predicted fate—for example, women who have been attested a high risk of breast cancer: "…many of these women described their risk in absolute rather

---

[31] A few weeks after the session many clients cannot even remember the figures or their meaning. "The majority of participants had transformed the estimates they had been given into accounts that 'fit' with their experience" (Lock, 2009, p. 75). Hallowell and Richards (1997), who analysed a large number of empirical studies on risk perception after genetic counselling, thus suggest that comprehension of risk should not be the success criterion for the counselling session, which would then be seen as very ineffective.

[32] To be able to assess abstract risk numbers, experts as well as laypeople always refer them to threshold values that are understood as the benchmark for normalcy: "The numbers conveyed a message that no one ignored, although over time the actual figures seemed to be of no importance. What mattered was their position in relation to the boundaries of normalcy" (Adelswärd and Sachs, 1996, p. 1186). Pregnant women who, after a first trimester test, have been labelled at risk for a child with Down's syndrome interpret this risk with reference to the threshold value 1:300—even if this is purely conventional. If the attested risk is less, they are reassured. If, on the other hand, it is higher, they believe that something is wrong with them or their pregnancy, see Schwennesen, Koch and Svendsen (2009).

than probabilistic terms—they felt they would definitively develop cancer in the future" (Hallowell, 1999, p. 605). Patients who have been attested a risk often imagine themselves living on the edge of an abyss. They no longer feel healthy and think they are marching directly toward a frightening disease (Gifford, 1986; Kavanagh and Broom, 1998). Although they are in good health, a risk attestation transforms the body into a source of latent harm. For this reason a woman attested an elevated risk for ovarian cancer after a PAP test wants to have everything removed that is not necessary for her life: "Because the tiniest bit can go wrong, and if that's not there, well, you can't have a problem with it" (Kavanagh and Broom, 1998, p. 440).[33]

3.2.4. Life in irrealis mood

The example of counselling for pregnant women very clearly shows the pathogenic effects of medically attested risks. A pregnant woman finds herself in a particular condition: she is in "good hope" as a German expression beautifully catches it; she is expecting a child. This child is not yet born but is already latently present. To some extent the woman is carrying the future in her body.[34] In this particular condition, in which a becoming "you" is latently present, women are susceptible to anxiety. Until a few generations ago it was customary to take care of and protect pregnant women from anything that may frighten or cause them anxiety. Today, however, the opposite is true: in an especially sensitive life situation women are besieged

---

[33] As a matter of fact, preventive mutilation is a medical option for risk reduction nowadays. Women who have been attested an elevated risk for breast cancer because they have tested positive as BRCA1 or BRCA2 are recommended that they undergo the preventive amputation of ovaries and breasts, see Footnote 59.

[34] The manner in which women experienced their pregnancies before its scientification by biology and medicine has been very clearly described by Barbara Duden. See, among others, Duden (1991; 2000; 2002a).

by professional counsellors who convince them that they must become aware of risks and make informed decisions.

The example of Ms G. reveals how the lessons on genes and risks generate profound anxiety and neediness among women. Together with her husband she walks into the genetic counselling centre in S. on a sunny day in May. She appears to be uneasy. At the beginning of the session she announces there is a rumour in her family that a cousin had terminated a pregnancy after receiving a diagnosis of Down's syndrome. This is why her gynaecologist had referred her to the genetic counselling centre. Ms G. is unsettled, and it is obvious she is hoping to hear some reassuring words. Her sweater clearly reveals a growing belly — she is already six months pregnant. The geneticist, who has worked for many years in genetic counselling, cannot give her what she needs. On legal grounds alone it is not possible for him to reassure his client. If after birth the child turned out not to be healthy, he would be guilty according to German law — guilty of the birth of a disabled child.[35] Therefore he confines himself to informing Ms G. about the eventualities, possibilities, and options. He reels off the normal information programme for pregnant women in order to prepare his client to make an "informed decision".

As in every genetic counselling session the geneticist first begins with a detailed questionnaire. The story of her cousin, how Ms G. heard about her from her mother-in-law, remains unknown to the geneticist. Since there was no chromosome analysis before or after the abortion he can only speculate. Later he will discuss chromosome aberrations in detail. Now he runs through his normal counselling programme for expectant mothers. He asks Ms G. about her pregnancy and looks through

---

[35] A geneticist at the same counselling centre says he fears the reproach: "Why didn't you tell me? And in Hanau there was a verdict... a thirty-four-year-old woman sued a gynaecologist because her child had Down's syndrome. She claimed she had not been forewarned. And the gynaecologist was forced to pay child support" (Samerski, 2002, p. 222). See also section 2.3.3.

her maternity records. Subsequently he writes down her family tree. So that they won't overlook any potential risks, he says, he is searching for any abnormalities among the couple's relatives. When Ms G. mentions that her nephew and her mother are hard of hearing, he digs deeper. He wants to know if the hearing impairment has existed since birth, whether the nephew wears hearing aids, when it was diagnosed, and takes careful notes. The geneticist also sits up and takes notice when he hears about her mother's miscarriage, but then admits that "the frequency... tells us nothing". After failing to find any other "abnormalities" he ascertains that apparently no additional risk factors exist. Now he takes the data he has gathered about Ms G. and constructs a risk profile: 31 years old, in her 23rd week of pregnancy, cousin with an abortion for unknown reasons, suspicion of Trisomy 21, nephew with hearing impairment. This risk profile or statistical double serves as the basis for the rest of the counselling session: from these traits the geneticist derives statistical probabilities which he then assigns to his client as something personal and meaningful.

After the preliminary questionnaire the geneticist moves on to the educational part of the session. He lets the prospective parents know that he will now lay out for them the basis for their informed decision.

> C: I now want to dig a bit further with this problem. So that I can better explain to you what chromosomes may have to do with it. And it is always part of what we do to create a basis for understanding and for dealing with the problem. I don't want to scare you now. This is simply part of the process.

Couple G. learn that with childbearing all kinds of things can go wrong. From the medical perspective alone, the fact that Ms G. is pregnant encumbers her with risks, and the geneticist lists the various risks Ms G. is taking with her pregnancy.[36] Her child, he

---

[36] From the medical perspective every pregnancy without exception is risky. In Germany, approximately three out of four pregnancies today

states, could have not only Down's syndrome but also a whole range of other diseases and disabilities. Since she is pregnant, he explains, she carries a so-called "base risk that something might be wrong with the child". Then he advises his client to forget this risk.

> C: If people thought about it (...) no more children would be born. Obviously everyone wants a healthy child.

He then continues with his explanation: this base risk is 3.5 percent. He lists the diseases and deformities that fall under this base risk: "chromosome abnormalities" and hereditary diseases such as "cystic fibrosis", "dwarfism", "bone growth disorders", and "albinism", and "hearing impairment". In addition, there are other multifactorial diseases such as "cleft lip and palate", "congenital heart defect", and "spina bifida". In detail he informs the prospective parents about everything that could go wrong with their child. Then he starts talking about a particular risk: the risk of chromosome abnormalities. The prospective parents learn about chromosome behaviour during gamete formation and cell division. In the process, the geneticist tells them, something can go wrong. When three instead of two chromosomes 21 land in one of her egg cells, then the child could have Trisomy 21 or Down's syndrome. "And this could happen to any of us", he comments. But it happens more often when women are older. He names a few figures and explains that after 35 years of age the risk increases "quite steeply". For this reason women over 35 are offered amniocentesis. However, more and more young women are choosing to be tested as well. Her own risk is "1 to 862", the counsellor states, as he jots down the figure.

> C: There are certainly young parents who can be affected by this because they are the ones less likely to undergo amniocentesis.

---

are expressly classified as "high-risk pregnancies", see Schwarz and Schücking (2004).

Then he points to the miscarriage risk and clarifies: "But everyone decides for themselves and receives the test if they want." In rare cases, the counsellor continues to point out reasons for concern, Trisomy 21 can be inherited—namely when two chromosomes adhere together, a so-called translocation. He now explains in detail the heredity processes when a translocation is present and the possible implications for the fetus.

He would not assume that something like this exists just on the basis of the story about her cousin—"but it could exist", he adds. "It's reported in books." Only a test can rule it out. If they want to be certain, they can have her chromosomes tested. The child could still have Trisomy 21, Down's syndrome, but not the hereditary kind. But if they want to really be sure that the child does not have any chromosome abnormalities, then she must undergo an amniocentesis or cordocentesis. However, these tests carry the risk of provoking a miscarriage. Amniocentesis would deliver results only in her seventh month. Moreover, there is no cure if the findings are positive. "What if something is found, what then?", he asks them to consider. And clarifies: "I have to leave the decision to you. I can only inform you." The geneticist then returns to the hearing impairment. He explains that this is an "autosomal recessive" inheritance and that "several genes" exist that can make "a hearing impairment". If both parents carry the gene, the probability of recurrence is 25 percent. But since the impairment occurred in a distant relative, it would not be likely to recur—but "it can happen to any of us because we carry in us one to ten of such mutated genes". For the unborn child he finally calculates a recurrence probability of one-sixteenth and calls this "low".

The session with the geneticist has not reassured Ms G. On the contrary, instead of calming her fears he has given her more reasons for concern. Toward the end of the session Ms G. looks very distressed and insists on chromosome testing for herself and her husband. Unlike his wife, Mr G. has an optimistic disposition and came to the counselling session brimming with confidence. But as the session continues he begins looking more despondent. When the counsellor briefly steps into an adjacent room and I show Ms G. where the restroom is, he lets out a deep sigh: "Huh, man-o-man-o-man!", he exclaims, sounding stunned.

During the session Ms G. does not learn anything about herself or her unborn child. There are no indications that something might be wrong with the child, only that the geneticist cannot rule anything out. And he speculated for almost an hour over what cannot be ruled out—expanding the horizon of alarming future possibilities.[37] His speech was interspersed with sentences focusing on "could" or "would". This manner of expression, talking about speculative possibilities, is called the irrealis mood, or subjunctive mood, in grammar. It identifies what is only imagined, speculative, and fanciful. In the genetic counselling session this declarative form is very dominant. As soon as geneticists apply their biostatistical expertise to their clients, that is, treat them as risk profiles, they start talking in the irrealis mood: they speculate over what could happen and corroborate this "could" with probability figures and risk curves.

The geneticist speaks not only in irrealis mood but also places his client in an irreal state. He asks her to become aware of the risks. But risk awareness fixes the gaze on a possible future and paralyses one's sense of the present. Clients are being asked to be where they are not and perhaps never will be. The Brothers Grimm tale of "Clever Elsie" tells of this anticipation of a speculative future and the paralysis in the present. Clever Elsie is a symbol for life in irrealis mood: she remains sitting in the cellar crying and paralysed under a walled-in pick-axe because this could kill her child who is not yet born. Upstairs,

---

[37] Human genetics, observed May and Holzinger (2003), produces uncertainty: human genetic research results in an epistemic paradox: it leads to an *increase of knowledge* and simultaneously *produces uncertainty* (May and Holzinger, 2003, p. 69). An interesting question, of course, would be what "knowledge" here means when it produces uncertainty. What form of knowledge, for example, are statistical probabilities from the perspective of a prospective mother?

Hans, her suitor, is waiting with her parents. Glancing at the pick-axe Elsie anticipated their future together and a possible misfortune: if she marries Hans and has children, and if she one day sends her child down to the cellar for beer, then it could be killed by the falling pick-axe. Pondering this menace, Elsie remains sitting and lamenting this possible fate.

As a pregnant woman now informed about genetics, Ms G. also lives under the sword of Damocles with a speculatively anticipated future. When she entered the counselling room she was unsettled by the rumour about her cousin. Now, as she leaves the room, her head is filled with the countless risks that could destroy her maternal happiness. The counsellor has asked her to imagine various alarming futures with her child. She is to envision that her child may have gene mutations, unknown deformities, and chromosome abnormalities in particular. Ms G. must reckon with carrying a future calamity inside her.

There are no tangible, present, and perceptible reasons for this. Whereas Clever Elsie let herself become paralysed by glancing at the pick-axe, which she imagined falling and killing her future child, the fear of risk has no concrete grounds. It is based on the construct of probabilities; its cause cannot be experienced or perceived, but is fully abstract. The fear of risk thus renders one helpless. That Ms G. wants to be tested is thus understandable. Those who live in the irrealis mood cannot stay in their right minds and become dependent on experts and technical equipment. "Good hope", transformed into the fear of risk, generates a boundless need for reassurance: for tests offering assurance that the evoked calamity — probably — is not yet there.

3.2.5. *The genetic risk*

If we add the many brief educational discussions taking place in gynaecological practices about deformities and pregnancy risks, then hundreds of thousands of women are placed in the irrealis mood every year, in a risk-laden state. Add to this a growing number of women and men who are neither expecting a child nor have anything else; they land in the irrealis mood because a geneticist has attributed a risk gene to them. Like Ms G. they discover they are carrying a future calamity inside themselves, although not in the form of an unborn child but in the form of a gene. What may come, they are informed, could already be present in their own body as a cancer gene or a genetic defect. In contrast to Ms G., whose duration in this state is limited to nine months, gene carriers remain in it their whole lives. Ms G. would have to learn that the gene error that genetic testing would likely bestow on her is part of her blueprint. The counsellor emphatically points out that the gene mutations that "cause" cancer are in every cell and have been there "from the beginning". Risks derived from genes thus have a particular efficacy: in contrast to other physical risks, such as the risk of cancer after a positive PAP test, a genetic risk is neither temporally nor spatially limited. It is always there and everywhere; it is the basis of one's own existence.

Indeed the geneticist admits the risk that she ascribes to Ms M. after genetic testing has a special status. She distinguishes it fundamentally from the risk that she derived from the family tree and the polyp finding. While she identifies the latter as an "assessment", she calls the disease probability after genetic testing a "concretization" of the risk. With this she is suggesting that the risk derived from the family tree was only speculative but the risk derived from genetic testing, on the other hand, is tangible and real:

> C: The test is really the <u>concretization</u> of the risk, of what at the moment is just an assessment.

Such formulations lead people to believe that a risk derived from DNA traits differs fundamentally from other risks. In another counselling session, the session with Ms K. about a breast cancer genetic test, she makes the same distinction more clear: the risk derived from the family tree is "only on paper", whereas Ms K. "has" the risk derived from genetic testing. With this distinction the geneticist promotes a grave misunderstanding: genetic risks, she claims, quantify something concrete, an inner menace or physical susceptibility. She leads her clients to believe that genetic risks are not speculation but diagnoses. Yet, no matter from where a risk is derived, its validity always remains the same. Risks can differ quantitatively but not qualitatively. Family tree data and genetic data function in the same way as the traits which are used to place clients in statistical populations. Genetic risk can therefore say nothing tangible about Ms M. or Ms K. With genetic risk the geneticist is only ascribing to her client a statistical probability. This is not concrete; nor can a person "have" it like an ulcer in their stomach or a cavity in a tooth.

How contradictory this alleged "actual" or "concrete" genetic risk is that the test promises to unearth is manifest in the counselling session with Ms K. Ms. K. is in her early twenties and very healthy. Her father's sister developed breast cancer when she was young, as did his mother. This is enough for Ms K. to meet the international criteria for a high-risk person. Therefore she now belongs to a population with an elevated probability that the outcome of genetic testing will be positive. The counsellor immediately offers her the option of the test. She explains what findings the test can yield and what these would mean. If the findings were negative, Ms K. would be released from the risk group and would belong to the average population with the average risk.[38] The test would not exclude the possibility of breast cancer; she could still develop it, like every other woman.[39] If the findings were positive, she would be placed in a population with an elevated risk:

C: So when a mutation in BRCA1 or 2 is carried.

---

[38] With breast cancer genetic testing in Germany a relative with the disease is always tested first. If the relative tests positive for a genetic mutation, only then will the actual test candidate be tested. If, however, the test of the relative with the disease turns out to be negative, as is the case here with Ms K.'s cousin, then geneticists assume that the disease is still genetic, but the corresponding gene has not yet been discovered. The candidate herself is then not tested at all.

[39] If Ms K. is not considered to be a gene carrier and still develops breast cancer, then her disease would be classified differently. Elsewhere the counsellor states: "If, let's say, they found something with your cousin and not with you, and you still developed breast cancer, then we would say it is not hereditary."

W: Yes

C: Then a woman who carries this mutation, from a statistical perspective, which says nothing at a personal level, has lifelong, an approximately 80–85 percent risk of developing breast cancer.

Recent studies have invalidated the figure of 80 to 85 percent as inflated.[40] Depending on who conducts a statistical study, how, and on whom, the risk figures vary considerably (Begg, 2002). Therefore, by 2003, the German Medical Association was already stating the disease probability with a wider range, citing 40 to 80 percent (Bundesärztekammer, 2003, A 1298). The American Cancer Society indicates 55–65 percent for BRCA 1 and 45 percent for BRCA 2.[41] Nevertheless, apart from the size and validity of the statistical correlation, the geneticist here expressly points out that a purported "concrete" and "actual" genetic risk says nothing about a person. She concedes that even after genetic testing it will remain unclear what will happen to Ms K. The genetic finding would merely be a new marker that would modify her risk profile. This new risk profile would allow the geneticist to make the following statement: if Ms K. had 100 lives, then in 80 of these lives she would develop breast cancer before the age of 79, and in 20 lives she would not.[42] In reality, however, Ms K. has only one life. What happens to her in this one life — the only question meaningful for Ms K. — is still written in the stars.

---

[40] Chen and Parmigiani (2007).
[41] See http://www.cancer.org/cancer/breastcancer/detailedguide/breast-cancer-risk-factors [7.8.2014]. Meanwhile various mathematical models and computer programs are being tested that calculate family histories in addition to genotypical markers and conduct a multivariate risk assessment. See Euhus (2001), Fasching et al. (2007). The goal is to derive the risk of disease not only from the BRCA1 and 2 gene but rather to generate an "individual" risk profile on the basis of various genotypical and phenotypical traits.
[42] Many studies take as their basis a life expectancy of 79 years.

The next sentence reveals the entire inconsistency of alleged "personal risks". The epistemic confusion between person and risk profile leads to downright absurd statements: although the counsellor just clearly stated that the genetic risk is "from a statistical perspective" and "says nothing at a personal level", she regards Ms K. as in danger. She warns Ms K. to be "careful" and urges her to undergo regular screening:

> C: So these are very high figures,
> W: Hm=hm
> C: Which means you must be careful. And that is also the reason why we recommend (-) early detection screening at close intervals.

In only a few sentences the geneticist has uttered very contradictory statements. She clarifies that the genetic risk says nothing about Ms K. — yet then she claims she is in danger. Such inconsistencies and absurd statements are typical in genetic counselling. In another counselling session, for example, a geneticist ascribes to her pregnant client a very low risk of bearing a child with cystic fibrosis. When the client is about to lean back, reassured, she adds: "But if it happens to you, it will be 100 percent."

Obviously not only laypeople but also experts suffer from epistemic confusion. They mistake the statistical construct about which they are making statements with the concrete person to whom they are speaking. Although the geneticist is aware of how meaningless a probability is for an individual case, she still considers her client to be in danger. Obviously she is confounding her clinical thinking that deals with concrete persons with the logic of populations and probabilities. Ms K. is not "suspect" because of any indices or concrete evidence, but, as is the case with a preventive grid search, solely because of her risk profile. She has traits that correspond to the search grid used to generate a target group of possible "perpetrators" — here affected persons. Similar to how the trait grid "male, age 18 to

40, (former) student, Muslim, Middle East country of origin"[43] turns people into "not-yet terrorists", the trait combination "female, early 20s, breast cancer twice on paternal side under age 40" transforms Ms K. into a "not-yet cancer patient". A genetic test would not change this abstraction; the finding BRCA1 or 2 positive would only add another variable to this profile, a variable that significantly diminishes the circle of preventive suspects. Not a clinical symptom, a physical susceptibility, or a diagnosis is the basis for the medical suspicion of cancer, but the mere fact that Ms K. belongs to a risk group based on traits.

The geneticist therefore warns Ms K. to be "careful", because her risk profile corresponds to the search profile for people who have not yet developed cancer. The warning to "be careful", however, does not mean that Ms K. could actually do anything to evade any concrete dangers—after all, she is not being threatened by anything concrete. Here the warning "be careful" means to step into the irrealis mood. Starting now, as a potential gene carrier, Ms K. is to spend her life in an artificial "not yet". Every six months, her counsellor emphatically recommends that she undergo an ultrasound[44] examination of her breast and abdomen to make sure that—probably—nothing has yet happened.[45] A positive genetic test would embody this irrealis mood for her: despite being in good health she must live with the feeling of carrying around inside her the preliminary stage

---

[43] In April 2006 the German Federal Constitutional Court banned the use of preventive search grids without an imminent threat because they violate the right to informational self-determination, see http://www.bundesverfassungsgericht.de/pressemitteilungen/bvg06-040.html [7.8.2014]. After September 11, 2001, different countries have conducted dragnet investigations for so-called "sleepers".

[44] Ms K. has swelling and lumps in her breast and thus frequently visits her physician. Mammography, they tell her, is not informative enough in her case. Therefore the geneticist recommends she undergo an ultrasound.

[45] Test results cannot guarantee health. It would be misleading to hope to gain certainty from test results. The statement of nothing positive can only be found for the moment and with a certain probability.

of a frightening disease. She will no longer consider herself healthy but as not yet sick.

Ms M. also learns what it means to live in the irrealis mood. She is taught that as a gene carrier not only would the feared colon cancer be lurking inside her but so are numerous other possible diseases. The gene mutation that she probably has turns her body into the subject and object of an all-encompassing and incomprehensible menace. After the geneticist explains to Ms M. about mutations and genetic testing, she reveals to her that a mutated gene can not only "trigger" colon cancer but also all kinds of other cancer diseases. The "colon cancer-associated gene" bestows on her not only an elevated colon cancer risk but also a host of other risks—each one scarier than the other. Stomach, ovaries, urinary tracts—her whole body is risk-laden, would turn into an alarming "not yet":

> C: And then there are other cancer diseases that have been found with these families. Including um... <u>possibly</u> an increased incidence of stomach cancer ((Ms M. rolls her eyes)).
>
> [...]
>
> C: Um, (--) then um (-) there can also be an increased incidence of ovarian cancer ((Ms M. leans forward, her eyes big)), which is also, which is also not uncommon, so it's <u>very important</u> for you to participate in gynaecologic screening and early detection.
>
> [...]
>
> C: and also ovarian, and this is a cancer disease that occurs more frequently in comparison to the average population.
>
> [...]
>
> C: No? Then um, there is also an increased incidence of cancer in the urinary tracts (Ms M. purses her forehead and raises her eyebrows)).
>
> [...]
>
> C: Then there is still an increased incidence of... um cancer diseases in the area of the pancreas, the bile duct, and so... for early detection screening you should have an abdominal ultrasound exam every year.

Genetics and statistics create a new "divinatory space", as described by anthropologist Margareth Lock (1998, p. 9). The reason for action is a new perception of the body: the body as a time bomb. If a gene mutation is eventually ascribed to Ms M., she will no longer be able to feel comfortable in her own skin. Malignancies and disease would be threatening her everywhere. Even now, as a possible gene carrier, her counsellor advises her to undergo screening at the recommended intervals. Ms M., who was originally afraid of colon cancer, must now be checked on a regular basis to see if she may have stomach cancer, uterine cancer, ovarian cancer, urethral cancer, or pancreatic cancer.

3.2.6. The genetic self

The anthropologist Emily Martin examines how social relationships are mirrored in our understanding of the body. In her monograph *The Woman in the Body* she analyses how the basic concepts and images of the industrial society, for instance, "production" and "hierarchy", determine the scientific view of the female body — and ultimately also the manner in which women experience themselves (Martin, 1989). How this understanding of the body has undergone a radical shift in the second half of the twentieth century is the focus of her second book, *Flexible Bodies* (Martin, 1994). In an age in which self-responsibility, flexibility, and self-management have advanced to become our guiding principles, the body is increasingly described in corresponding images: the solid, machine-like, and hierarchically controlled body of the industrial age has been transformed into the flexible, open, and threatened body of the systems age. The epitome of this new body image is the immune system: a dynamic system that must continually adapt to its environment. It is never fixed and stable, instead always open and in a state of flux. Thus the immune system requires constant monitoring and optimizing — in the same way that modern workers must always manage and optimize themselves to satisfy the demands of the new corporate world. In an unsettling way social and economic demands coincide with the scientifically transmitted view of the body: "People are, so to speak, coming to see themselves as mini-corporations, collections as assets that each person must continually invest in, nurture, manage, and develop" (Martin, 1994, p. 77).

In the 1990s the "gene" substituted the "immune system". Today it is first and foremost genes that redefine the body as a risk-laden system in need of monitoring. How closely the genetic body corresponds to the social ideologemes of the twenty-first century is illustrated in a brief dialogue from a colon cancer counselling session with Ms S. The geneticist is teaching her client, who, like Ms M., came to the genetic counselling centre because a relative developed colon cancer, about the consequences of a potential gene mutation.

> C: This is a so-called mismatch repair, which means it is a function by which, when base exchanges in... with the, uh, with the reproduction of genetic material, as continually happens as cells grow, when random mistakes occasionally happen, this corrects them.
> W: Hm=hm
> C: There are specific protein molecules in the body that are in charge of this.
> W: Like the police
> C: Exactly, they are like the police. And this is ex... this is exactly the function of these genes.
> W: Hm=hm.
> C: And um --- and when they do not function right, then logically such changes can persist.
> W: Yes, hm=hm
> C: Other genes, which, on the other hand, like the tumour suppressor genes, are on guard to make sure cells do not further mutate. And when they, when they become nonfunctional, then cancer can develop.

The image of the police is prevalent. In other genetic counselling sessions geneticists themselves bring it up (Kovác and Frewer, 2009, pp. 215–16) and the Epigenome Network of Excellence, for instance, excessively uses criminological pictures for public relations. It describes cancer on its website as "Cancer is the 'enemy within', the criminal element that upsets the harmony of our body's cellular community. Our internal police force, our immune system, does all within its powers to hunt down and disarm these troublemaking cells" (Epigenome NoE, 2008a). The short texts are illustrated with pictures depicting policemen with truncheons and a man in handcuffs.

Most likely Ms S. has already learned from magazines and TV about the police in the cell nucleus. She has incorporated the pop science images of scientists who envision the body in analogy to the modern surveillance state and claim that "our body is under silent surveillance by a patrolling police force" (Epigenome NoE, 2008b). Now Ms S. is picturing her own body as if it were a modern surveillance state; she believes inside her there is a genetic police system that is "on guard", that is constantly on the hunt for any threatening breakdown and malfunction, clears out the problem, thereby stabilizing her health. Otherwise her body would get out of control. The big enemy, these explanations suggest, comes from inside: genes are like "terrorists", biologist John Turner writes: "They have the power to kill, maim, or make life downright miserable for us and our children. Some strike at birth, others 'sleep' for decades, and, like good terrorists, they are so well integrated into our body politic that, until the last few years, their exact whereabouts were a mystery: their individual extirpation (or more properly correction) is still well nigh impossible" (Turner, 2001, p. 8).

The medical sociologist David Armstrong maintains the emergence of risk or surveillance medicine is far-reaching because the "surveillance machinery" assigns people a radically new self-concept, which he calls a "risk identity" (Armstrong, 1995, p. 405). This new "identity" stands in the shadow of a risk-laden future. It is no longer, as in the age of clinical medicine, derived from a concrete body, but from statistical collectives and possibility spaces. "Surveillance analyses a four-dimensional space in which a temporal axis is joined to the living density of corporal volume" (Armstrong, 1995, p. 402).[46]

---

[46] On the significance of time in a future-orientated medicine, and on the change in the understanding of disease, see also Greco (1993).

Such a "risk identity", that is, an unsafe, precarious body, in which lurk imminent threats and ominous futures, is what a geneticist assigns her client. Ms M., Ms G., or in the above case, Ms S. learns to perceive herself no longer as a concrete person with a unique biography and open future but as the embodiment of possibilities and risks. Her existence as a gene carrier is no longer rooted in an experienced past and present but is derived from a statistically anticipated future. The enemy, genetic education suggests, comes from inside; future calamities lurk as faulty genes in your interior. This reinterpretation makes it seem plausible for Ms S. to accept the laws of statistical cohorts and probability distributions as her own. The attributed risks, counted for populations, seem to measure the threats from within. The genetic self she is asked to incorporate makes Ms S. more or less physically compatible with population-related risk calculations and prevention strategies.

The counselling clients cannot question their epistemic transformation. As I have analysed in section 3.1.1, geneticists claim to be experts on the human species and, consequently, do not talk to their clients as peers but as objects of their scientifically validated knowledge. Thus, the setting of genetic education does not allow clients to protest, doubt, counter with their own experiences, or form their own opinions. What is conveyed here as "knowledge" and a new self-concept is authoritative and non-discursive. It does not relate to everyday experience, and thus allows no appeal to common sense. The geneticist has therefore already provided the most important aspect of the session: the conceptual and ideological realm within which the client is to view herself and the decision she is called on to make. What can still be called a "decision" after receiving such a lesson in genetic self-perception is the topic of the following sections.

3.3. The Compulsion to Risk Management: The Decision

The geneticist lectures on genes and risks for 45 minutes to teach Ms M. the self-perception necessary for her to make an autonomous decision from a professional perspective. Now she turns to the genetic test, the actual reason for the counselling session. The goal of the session is to enable Ms M. to make an "informed choice" about this test.

"You belong to the group of high-risk persons", is the main statement the geneticist has made about her client. This risk attestation is the reason why Ms M. is a potential candidate for genetic testing and should also serve as the basis for her "informed decision". The geneticist says nothing about what she thinks Ms M.'s decision should be. The expert is very careful not to recommend the test to her client in any way. On the contrary, in response to the question of whether the test is advisable or not, she takes no stand. She wants the client to see the test as a mere offer. It is Ms M.'s task, clarifies the geneticist repeatedly, to make her own decision. To make it clear that the test is only an option but not an invitation or even a recommendation, she distances herself from the testing offer by throwing in the word "theoretically":

> C: You belong to the group of high-risk persons [...] and theoretically we can offer you this molecular genetic testing.

If someone declares an opportunity as "theoretically" possible, she is implying that this says nothing yet about the practical implementation. Often such a formulation is followed by a "but", that is, an objection that speaks against this possibility. In this sense, the geneticist is also trying to distance herself from her own offer: she wants to make it clear that the test offer involves a "can", not a "should". "We can only say what can be done but not what should be done", says another counsellor. During the rest of the session the geneticist repeatedly emphasizes that Ms M. herself must know what she wants to do. The test, and thus also the "interview" to prepare her for it, will be conducted only upon her explicit request:

> C: This means (-) ((very emphatically:)) <u>if you want it</u>, and you should think seriously about this, and then there will be another interview.
> W: Yes, hm=hm.

3.3.1. The imperative of the autonomous decision

At first glance the request for a client to make their own informed choice appears as emancipation from paternalistic authority and empowerment of the client. Just a few decades ago the demand to let patients choose was a provocation. Physicians assumed patients were not at all in the position to make their own decisions about health-related issues, and therefore expected that their instructions would be followed without protest. The expression "Doctor knows best" captures the unquestioned authority of physicians in health matters. The right to be informed about medical treatment had to be fiercely fought for by patient associations and the women's movement: in the early 1970s a first battle in the USA was won against the indignant medical profession following vehement public debates when "the pill" became the first pharmaceutical drug to come with a package insert for consumers. However, what physicians bitterly fought as an invasion in the physician-patient relationship soon became very convenient: if they had proof of their patient's informed consent, then they were safe from possible lawsuits. They no longer had to bear responsibility for the risk of harming a patient's health with the pill, but could shift the responsibility to their patients (Watkins, 1998). A few years later, in 1976, the Supreme Court judicially dethroned the medical experts: in a landmark decision the court decided that the decision about life-prolonging measures should be made not by physicians but by the patient's closest kin. They granted the parents of a young woman lying hopelessly in a coma the right to have their daughter taken off artificial respiration against the advice of her physicians.[47] In Germany, however, it was not until the late 1980s that "informed consent" became a universal patient right and entered into routine medical practice. Here it is first and foremost a ritual to protect

---

[47] David Rothman (1991) describes the shift from the dominant authority of medical experts to formalized and bioethic decision making at the patient's bedside in the USA.

physicians from claims of damage: patients have to sign that they accept the risk of waking up after anaesthesia with a punctured colon or not waking up at all.

Meanwhile, the right to make an "informed choice" has largely lost its original intent of protecting citizens from medical transgressions. Making choices today is not a freedom; rather it has become a new obligation. The modern person, according to sociologists Ulrich Beck and Elisabeth Beck-Gernsheim, has become *Homo optionis*. Whether we are talking about gender roles, faith, or genetic testing options: an autonomous individual today is the person who chooses. Caesarean or home birth? Alternative school or traditional school? Single or family? Buddha, Allah, or God? Active or passive euthanasia? For every life situation the "decision society" (Schimank, 2005) has options available, and the modern person must do just one thing: choose.

The call for autonomous decision making is especially prevalent in health care. Pregnant women learn to assess the risks of Caesarean and vaginal birth; cancer patients learn to assess the success rates of chemotherapy, radiation therapy, and surgery; and the palliative physician gives his patient the options "euthanasia" or "continue to live". William R. Arney and Bernard J. Bergen (1984) have written the history of this inclusion of patients in medical practice: not, as is generally the case, as emancipation from medical paternalism but as the history of expanding medicalization. As the authority of the physician diminishes, his realm of competence grows. The expert loses his hegemony and becomes a facilitator, in other words, an "enabler" and "supporter" of decision making. At the same time medicine targets no longer just the sick but also the healthy—inclusive of their familial, psychological, and social context. Medicine therefore no longer limits itself to the pathological but has extended its arm to manage people comprehensively over their entire lifespan and to advise them to practice self-management. "Medicine became an integrated organization that extended outside the boundaries of the hospital and managed patients instead of treating diseases. The organization would be regulated not in accordance with normality but in accordance with social utility and would incorporate the anomaly by creating him or her as the 'chronic patient' in need of total care" (Arney and Bergen, 1984, p. 80).

Genetic counselling is by large paradigmatic of these efforts to educate people to become managers and decision makers of their own affairs. Genetic counsellors expressly see themselves as decision making facilitators; their service has the goal of producing an "informed decision". Indeed, geneticists pursue this goal not only on paper but actually in practice. In each counselling session I have observed the counsellors repeatedly insist that clients must make their own decision — and alone bear responsibility for the consequences: "You must make the decision, because we don't bear the consequences", one counsellor repeatedly hammered home. Another counsellor made it clear right at the beginning of the session: "Advice you will have to seek... within yourself."

3.3.2. The option requiring a decision: The test

Ms M. sought out the genetic counselling centre because she wanted to do something preventive against colon cancer and has received training in being an autonomous decision maker for matters involving genes and risks. The geneticist has ascribed to her a risk profile and has taught her how to make an informed decision about the option of genetic testing. When she moves on in the counselling session to talk about the test thirty minutes later, she mainly explains the genetic background. She describes the procedure, including the necessary preliminary investigation of the polyp tissue, names the genes that can be identified, and points out that it is possible that the test may deliver no conclusive results because no known gene mutations are found.

First Ms M. learns that her family's cancer is classified as Lynch Syndrome, or HNPCC for hereditary nonpolyposal colorectal cancer, and that the involved genes are known:

> C: And there are genes, these are just names now, which don't tell you much, they are MSH2, MLH1, as well as a few other genes that can be identified around 60, 70 percent of the time.

Here the geneticist names the genes that are said to be critical for Ms M.'s health. What for Ms M. is an impalpable menace seems so clear and tangible to the expert that she can name them. The geneticist, however, mentions only abbreviations, that is, she refers to them, and then remarks that this information does not tell Ms M. much anyway.

For the test, the counsellor explains, the polyp tissue from Ms M. would first be genetically analysed. They would look to see whether the genes show any "mutation" and whether any function does not work as it "normally" does:

> C: In fact, these genes have, which I showed you here (leafs), have in common a certain characteristic, a so-called microsatellite instability. This is (-) an unfamiliar term.
> W: Yes, hm=hm
> C: This is simply a typical, uh, function that these... which the proteins associated with these genes carry out. And if there is a mutation in one of these genes, then this function doesn't work as it normally does.

Here as well the geneticist makes references that Ms M. cannot follow. She suggests that something may have mutated inside Ms M. and is no longer functioning normally. Ultimately Ms M. can only understand from this that she may not be genetically normal and that geneticists are able to find this out.

If the microsatellite instability is found in the polyp tissue, which would be a sign of a gene mutation, then they would draw blood from Ms M. to isolate DNA and

> C: ...then very specifically analyse these genes. This means, in other words, they will go through these genes and look at each base to see if everything is normal or if there is a mutation anywhere that can cause disease.

Geneticists can also look, this formulation suggests, to see whether Ms M. has an anomaly inside her (see section 3.1.4).

When the counsellor subsequently declares that the test would "concretize" the risk, it seems as if the risk would quantify Ms M.'s health condition, that is, as an incongruity or a malfunction on the genetic level.[48] In reality, however, the test can only reveal a genetic trait that classifies the person being tested. Genetic testing can "differentiate" tested family members, states the geneticist, and place them in two groups: those with an average risk and those with a high risk. The latter group is then offered "qualified" early detection.

The geneticist has now described the test from her perspective. She has represented it as an opportunity to pinpoint disease causes and anomalies and improve preventive care—namely as the way to offer clarification, knowledge, and a course of action. She does not give any reasons for not doing the test. At one point she remarks that some people do not want the test. But she says nothing about how they justified this "no", and if their decision eventually turned out to be pragmatic or judicious. Nor will she recommend the test. Several times she points out the option of not being tested. Ms M. can withdraw at any time from the testing procedure, she clarifies. When she mentions that Ms M.'s sons can be tested only after they are of age, she emphasizes again that the decision is a very personal one:

> C: They must first consider whether they even want it. There are many people, such as you heard from your sister ((Ms M. raises her eyebrows)) whose eldest said she didn't want to be tested.
> W: Yes, yes.
> C: And who knows, perhaps the children are... Everyone is different, and we don't want to push anyone into testing, but we really want to work together with you to find the right path.

---

[48] In 30 percent of cases with HNPCC syndrome, however, nothing is found, continues the geneticist: "They simply look at the genes and find nothing." For geneticists, however, this does not mean that the disease is not genetically caused—the corresponding gene is just not yet known: "We just don't know all the genes yet."

Ms M. is certain that she wants to undergo genetic testing:

> W: By all means I want ((gestures with her arms)) uh the (--) to know about, uh, the genes.

Ms M. has no doubt. "By all means", she says, emphasizing her resolve. The other clients whose sessions I observed also firmly stated their desire to be tested. Ms R. does not want to remain "naïve and ignorant", and Ms S. exclaims that she does not want to stick "her head in the sand". These reactions are not surprising: even though the counsellors insist on clients making their own decisions—the explanation about allegedly personal risks and genetic testing options unavoidably produces thought and action constraints. It is common practice among genetic counsellors that their explanations blaze a trail to the test, but they strongly emphasize that their clients must make their own decision. "While the autonomy of the individual or the family is often expressly valued and respectfully mentioned by the counsellor, the dialogue also makes clear that a genetic test creates an avenue to make a knowledge-based choice of prevention" (Koch and Svendsen, 2005, p. 828). In interviews women recount how they felt pressured by the counselling. The promise of feasibility and decidability made them responsible for preventing the disease. Their fate suddenly seemed to be placed in their hands (Hallowell, 1999). The women, here burdened with an elevated risk for breast cancer, felt strongly obligated to do everything possible to not just yield to their fate. Not one of them had the feeling that they could freely make a decision. Every one of them felt that they had to submit to testing and other interventions: "For most, obtaining risk information and managing cancer risk was not seen as a matter of choice, but as a matter of necessity" (Hallowell, 1999, p. 613).

Ms M. hopes to gain information from the test and thus a new approach to fighting the feared disease. She believes she is doing something, and for her the test is the first step in this direction:

> W: Yes. That is, by all means I want ((gestures with her arms)) uh the (--) to know about, uh, the genes.

> C: Yes
> W: That is, whether the
> C: Whether that's the case.
> W: Whether that's the case, um… yes! That's interesting for me, and then, uh, (--) to what extent it can be contained and (-), that is (--), what more can I do about it besides maybe every year… and if then they can say, uh (--) if they can say more precisely what kind of cancer it is and (--) how I… What I myself can do, that is, one says, okay, now everything healthy has been removed, but if it comes back somewhere, whether a bit of colon is removed or something, that is; I'm interested then, so to speak, in what comes next, that is, if that's really the case, what happens next, what can I do.

For Ms M.'s early detection programme it makes no difference whether she has been tested or not. Every vigilant internist would recommend a regular colonoscopy because of her family history and polyp. The geneticist also suggests she have a colonoscopy every year. But this is not enough for Ms M.: she hopes the test will give her concrete knowledge that will enable her to proceed actively on her own:

> W: What then, what comes next for me, what can I do or what should I do, should I stop taking hormones or what. […] And if there's something I… I'm a little overweight, is that a problem? In other words, everything that (-) um, where can I start, what can I do.

Ms M. wants the test to tell her something about herself so that she will know how to deal with the imminent cancer disease. Apparently she believes that the "genes" will reveal something significant about her health and her future—a kind of inner truth that has hitherto been hidden from her. She assumes that the genetic test will reveal something tangible, even physical. This is how the geneticist portrayed the test: as a way that clarifies and reveals hidden knowledge.

But the hope that the genetic test will reveal new insights about her personal condition is deceptive. The genetic test does not enable a diagnosis by means of which the physical constitution can be systematically targeted, for example, through nutrition or movement—like a physician can diagnose celiac disease, for instance, and recommend abstinence from gluten-containing grain. What the geneticist can tell her based on the findings of the genetic test is ultimately not much: the genetic test will not tell her anything she has not already heard from her mother. Her mother always warned her, says Ms M.: "Hello, everyone in our family has died of colon cancer, so be careful." Neither the genetic test nor the counselling session will give Ms M. any new information in this regard. Ms M. has already had a polyp. She knows that she could develop colon cancer and therefore must be "careful". More than this neither the geneticist nor the test can disclose.[49]

Yet on another level the genetic counselling says and asks for much more than her mother's warning. The geneticist has objectified as a genetic risk the fact that Ms M. could become ill and has ascribed to her something physical. Ms M. must now envision that she carries what could happen inside her as a risk-laden gene. This makes it seem as if Ms M.'s fate is tangible, foreseeable, and decidable. The genetic risk suggests Ms M. could take control of her future—by being informed, tested, and monitored. This promise burdens Ms M. with a new responsibility: the responsibility for what could happen to her. Hallowell, who interviewed numerous breast cancer risk carriers, thus came to the conclusion that genetic counselling does not enable "autonomous decision making" but rather imposes on women new responsibilities and obligations. The responsibility to maintain their health and the obligation to manage their risks: "It can be argued that by labelling individuals as 'at risk', and presenting genetic risks as manageable,

---

[49] Since Ms M. has already had a polyp, it is unlikely that the test will give her the "all clear": and even if for some reason it turns up negative, it would still be medically judicious to continue being cautious and to act on the colon results and not the genetic test results.

genetic counselling implicitly places individuals under an obligation to modify these risks" (Hallowell, 1999, p. 599).

3.3.3. Self-determined helplessness

3.3.3.1. Obligatory risk management

The geneticist has portrayed the genetic test as meaningful, judicious, and helpful—but has also pointed out that refusing the test is also an acceptable option. If Ms M. does not undergo testing, the geneticist will not protest. For her the most important thing is that her client makes an "informed choice"— no matter what decision she makes. But the decision is only considered to be "informed" if Ms M. makes it within the framework she has been placed by the geneticist with her explanations. The goal of the genetic counselling session is "to put clients in the position to make their own decision about how they want to deal with their genetic risk" (Schmutzler *et al.*, 2003, p. 496). In any event, Ms M. must learn to see herself as a risk carrier and to feel responsible for "her" genetic risk. On this condition a "no" to the genetic test is acceptable—as an alternative path in risk management, as an analogous option to "yes". The litmus tests for the risk awareness of Ms M. are the recommended early detection tests. Whether her client is actually mature and self-determined, that is, responsibly copes with her genetic risk, will be apparent to the geneticist by her readiness to undergo oncological check-ups on a regular basis. She emphatically draws to Ms M.'s attention that her future depends upon her risk management. She formulates not only an imperative for decision making but also an imperative for risk management:[50] "Early detection", she exhorts Ms M., allows her to take her health into her own hands:

---

[50] Koch and Svendsen (2005) have also made this observation, that in cancer counselling sessions genetic counsellors demand from their clients both a "self-determined decision" on the one hand and "responsible", in other words medically prescribed, management of attested risks on the other hand. Decisions not serving preventive risk

C: I believe the most important, the very most important, thing you can basically do: early detection. That is, so to speak, what lies in your hands.

If the test is positive and Ms M. is labelled with a purported "real" risk, then from the perspective of the geneticist a colonoscopy would not only be a "should" but even a "must":

C: If the suspicion is confirmed, that is for hereditary forms, it must be done once a year.

For attested gene carriers monitoring is therefore not a matter of choice but an explicit necessity. However, this "must" is not a conventional medical "should", a medical indication. The geneticist is not basing her recommendation on symptoms or a diagnosis—for instance, Ms M.'s polyps, which can be considered a preliminary stage of a malignant tumour—but solely on her risk profile.[51] If she did not have any polyps, but would fulfil the risk criteria for other reasons, the geneticist would make the same recommendation. Symptoms or indications of Ms M.'s physical condition are not the basis for monitoring, but

---

management are not acceptable to the counsellors: "Thus, while the notion of choice is emphasised all the way through, some choices are more easily accepted than others. If counsellees take the decision in line with the ideology of prevention, these are accepted right away whereas decisions that do not lead in that direction are contested" (Koch and Svendsen, 2005, p. 829). They observed that geneticists oblige their clients to undertake risk prevention, and often genetic testing. In doing so, they appeal to their clients' "responsibility" for their own health and the health of their relatives. Koch and Svendsen talk here about a twofold imperative, the imperative of decision making and of prevention, see Koch and Svendsen (2005, p. 824).

[51] In risk medicine or "surveillance medicine", as Armstrong writes, disease symptoms are reframed as risk factors: "Symptoms and signs are only important for Surveillance Medicine to the extent that they can be re-read as risk factors." The geneticist therefore understands Ms M.'s polyp no longer as a symptom or sign of an "underlying" disease or anomaly, but as an omen: "It is no longer the symptom or sign pointing tantalizingly at the hidden pathological truth of disease, but the risk factor opening up a space of future illness potential" (Armstrong, 1995, p. 400).

only the fact that she has been classified by statistical traits as an at-risk person. The geneticist sees all the other risks she is ascribing to Ms M. as grounds for check-ups as well.[52] Ms M. should also, exhorts the geneticist, be screened regularly for all the other disease probabilities. An annual PAP smear, abdominal ultrasound, vaginal ultrasound, and an urine analysis will make sure that no tumours are growing in her uterus, ovaries, pancreas, bile duct, or urethra. The geneticist also mentions an annual gastroscopy, but she does not want to "prioritize this", because Ms M. has not yet been genetically tested.

Doing nothing in the age of risk management is regarded as irresponsible, even downright fatal. In medical genetics, declares the sociologist Thomas Lemke, self-determination "feeds back on social norms and material goals, which ensures an 'informed' self-determination, in other words, a specific practice of freedom" (Lemke, 2000, p. 253). And the goal is: self-responsible risk management. Therefore, Lemke continues, only those individuals who draw "the correct, that is, risk-minimizing and forward-looking, conclusions" are considered autonomous (Lemke, 2000, p. 252). Not complying with the recommended surveillance measures, whether due to aversion, faith in God, scepticism, or indifference, is downright fatal in the logic of risk medicine.[53] Risk carriers who do not actively manage their risk are regarded either as in need of counselling or as having consciously chosen their misfortune. Women placed in a high-risk group because of genetic testing are there-

---

[52] William R. Arney describes medicine in the second half of the twentieth century as the medicine of "monitoring", of ongoing surveillance, whose goal is preventive optimization: "The new medical logic proposes to meet disruption and dislocation with a policy of '*preventive optimization* and not alleviation' which extends over the entire life course", see Arney and Bergen (1984, p. 113).

[53] Women who do not regularly undergo the recommended check-ups have been excluded by the testing prevention programme of the German Consortium for Hereditary Breast and Ovarian Cancer which was financed by German Cancer Aid, see Gibbon, Kampriani and zur Nieden (2010, p. 457).

fore quick to declare they will do everything possible: "Many made clear during the interviews that they had done all they could do, or were currently prepared to do, to manage their cancer risk" (Hallowell *et al.*, 2004, p. 562).

A major function of prevention campaigns, according to the anthropologist Lisbeth Sachs, is to saddle citizens with the responsibility for their own health. Although they are increasingly less able to determine and shape their environment, they are being persuaded that they are responsible for whether they become ill or not. Illness is thus no longer a stroke of fate but the consequence and evidence of their incompetence — their incompetence in making the right, informed decisions:[54] "A health that can be chosen... testifies more than just a physical capacity, it is the visible sign of initiative, adaptability, balance and strength of will. In this sense, physical health has come to represent, for the neo-liberal individual who has 'chosen' it, an objective witness to his or her suitability to function as a free and rational agent" (Greco, 1993, pp. 369–70). This talk about decidability and responsibility mobilizes. Even if they are not convinced about the possibility to control their fate, women classified as high risk for breast cancer, after having been tested and counselled, feel "obliged to 'do something' about their health or feel guilty if they do not act" (Gibbon, Kampriani and zur Nieden, 2010, p. 460). The promise of feasibility keeps people in a constant state of fear of missing something and afterwards being to blame for their own misery.[55] This fear can be greater and more agonizing than the

---

[54] Prevention campaigns often portray diseases as the result of unhealthy decisions. Heart diseases, observe Davison and Frankel in their study on lay epidemiology, are regarded here as the result of ignorance or lack of self-discipline: "'Choosing health' is thus central to the official ideology, with the strong implication being that much heart disease is attributable either to ignorance or to a lack of self-discipline" (Davison, Frankel and Davey Smith, 1992, p. 3).

[55] Using the example of reproductive technologies, Sarah Franklin (1997) impressively describes how the promise of technical feasibility creates pressures and turns hopes and desires into needs and demands.

fear of the disease itself. When they can no longer regard their future as manipulable and trust their destiny, "high-risk" women can liberate themselves from their fear: "In the present study, by describing themselves as accepting whatever the future might bring, the majority of women presented themselves as able to contain or manage their fears about a future that they felt they could no longer manipulate and lead an anxiety-free existence" (Hallowell *et al.*, 2004, p. 563). But as long as their health is manageable, they can find no peace. If they become ill, they must blame themselves.

3.3.3.2. Mobilized helplessness

What at first glance may seem self-evident and rational, namely the striving for "knowledge", "prevention", and the corresponding decision making options, turns out on closer inspection to be highly irrational. The mobilization toward the risk management of their own affairs enmeshes clients in paralysing contradictions. What they had hoped for is not what the geneticist can offer them. Between their worries and desires on the one hand and the specialized knowledge of geneticists on the other hand lies a chasm. Ms M. is looking for personal assurance and receives risk figures. She expects knowledge that will enable her to act and receives laboratory results and statistical predictions. She wants to do something for her health and is called on to manage calculated risks. These contradictions and aporia to which women are led by the promise of "knowledge" and "decision" are very prominent in the counselling session with Ms K. (see also section 3.2.3).

Ms K., young and healthy, wants to be genetically tested. She hopes that the test will reassure her. As she herself says, she wants to "rule out some more" the chance that the breast cancer in her family "could be (-) hereditary". What she will face if the test does not give her the anticipated clearance is not discussed in the counselling session. Apparently, from the geneticist's perspective, learning about DNA sequences is a sufficient basis here for making an "informed decision". The geneticist is quite relieved when Ms K. agrees to postpone the topic of "positive test result":

> C: I don't know if we should continue talking about all the issues involved, what's possible, if a genetic mutation (-) is diagnosed.

Ms K. immediately agrees and again clearly expresses her hope that nothing will be found:

> W: I don't think we need to do that yet.
> C: Yes, I agree, we can do that...
> W: We can do it if there is anything at all.

Obviously the geneticist is not comfortable confronting a young and happy woman with the consequences of her services. If a "breast cancer gene" is ascribed to Ms K., everything will change. As a high-risk person without genetic testing results, Ms K. is already being asked to undergo regular check-ups—"just to be sure", the geneticist emphasizes: every six months a mammogram or ultrasound, breast palpation, an abdominal ultrasound. What now still appears as a preventive measure because something could happen or not would then become vital continuous monitoring. Genetic testing would turn the "could" into a genetically based "not yet". Her hope and assurance would abruptly change into a negative outlook. She would have to spend the rest of her life with the feeling that, despite being in the best of health, she is carrying a disease around inside her—a scary disease that could break out at any time. Furthermore, as with Ms M., she would be attested not only a genetic risk for breast cancer but also a whole host of additional risks. She would have to undergo yearly check-ups to see whether she has developed ovarian cancer, skin cancer, pancreatic cancer, or uterine cancer.[56]

---

[56] Many women have difficulties when they receive the results of genetic tests. For this reason they are often offered "psycho-oncological counselling", and the German Cancer Aid Consortium "Hereditary Breast and Ovarian Cancer" has created a checklist and a trait grid for

Nevertheless, despite the attested genes and risks, she would still not know what will happen to her. Whether she will become ill or not, and if so, when, at age 30 or at age 75—everything that is really meaningful in life, nobody can predict for her. Her future will remain just as open as before. Nor can she avoid the predicted calamity. The urgently recommended "early detection screening" makes it seem as if something can be done about the imminent disease. This promise is what makes genetic testing and risk predictions seem meaningful. Conversely, if nothing can be done, genetic testing forfeits its appeal: "If the future is perceived as unmanageable... then risk assessment (whether it is based upon one's lifestyle or one's genes) essentially loses its meaning or, at the very least, its perceived usefulness" (Hallowell *et al.*, 2004, p. 563).

The benefits of mammography are disputed.[57] The colonoscopy recommended to Ms M. can remove polyps before they mutate. Mammography, on the other hand, is not "preventive" care, but rather, as a client remarked, merely "follow-up care": it cannot prevent anything; it can only detect, and that only as a snapshot. Moreover, early detection is not always useful. "Earlier is not always better", says the National Women's Health Network, which through the insurance firm Techniker Krankenkasse published a brochure on the early detection of breast cancer (Nationales Netzwerk Frauengesundheit, 2007, p. 15). It is not clear how often early detection extends merely the time lived under the shadow of diagnosis but not life itself. For many women it might merely advance the time of diagnosis but

---

screening "psychosocial risk patients"—in other words, women attested with a higher risk of not coping well with the test results. See Schmutzler *et al.* (2003, pp. 499f).

[57] For more on the effectiveness of mammography as a population screening, see, among others, Gøtzsche and Olsen (2000), Gøtzsche (2003), Mühlhauser (2013), Nationales Netzwerk Frauengesundheit (2007), Swiss Medical Board (2013). For more on the effectiveness of early detection and other preventive measures such as hormone therapy, etc. for at-risk persons, see, among others, Calderon-Margalit and Paltiel (2004), Lux, Fasching and Beckmann (2006), Schmutzler *et al.* (2003).

does not postpone their time of death (Nationales Netzwerk Frauengesundheit, 2007; Rothman, 1998).[58] Mammography as a general screening misses every third to fifth tumour, produces a high amount of false positive results, and leads to overdiagnosis and overtreatment. Hence, there is no evidence for recommending it. Since incidence rates of cancer are higher in the population of so-called high-risk women, statistical cost–benefit analysis turns out differently here, but basic issues remain. Additionally, women at "genetic risk" are especially sensitive to radiation, and thus mammography can even cause breast cancer (Pijpe et al., 2012). There is only one way proven not to remove but to reduce the risk: prophylactic self-mutilation. Therefore the German Cancer Aid Consortium on Hereditary Breast and Ovarian Cancer expressly recommends that women who test positive have their ovaries and breasts surgically removed as a preventive measure (Steiner, Gadzicki and Schlegelberger, 2009, p. 28).[59] No wonder that women do not experience genetic testing and follow-up measures as the gain of control and autonomy as they have been promised to them (zur Nieden, 2013).

---

[58] This book is not a health guide. Whether women who undergo genetic testing and risk monitoring live longer than or not as long as others is not the issue here. I want to raise the question of the social and symbolic function of genetic education apart from the medical "body count". A sober glance at the statistics, however, can prevent one from being intimidated by the threats of disease and promises of health posited by risk medicine. It is mainly technology-minded physicians and misleading statisticians who fuel the rampant fear of cancer and the need for health surveillance. Preventive medicine is therefore an outstanding example of the irrationality of a society governed by the belief in technological feasibility.

[59] Approximately 1 in 10 women in Germany have their breasts amputated after positive genetic testing; in the USA it is approximately 1 in 4, and in a Rotterdam clinic 1 in 2 (figures from Steiner, Gadzicki and Schlegelberger, 2009, p. 29). Steiner, Gadzicki and Schlegelberger acknowledge, however, that the risk reduction here is only probable. Since no data from long-term studies exist, it is not possible to conclude that these risk-reducing operations offer a survival advantage (Steiner, Gadzicki and Schlegelberger, 2009, pp. 28–9).

3.3.4. Decision making: The paradox of personal risk assessment

For clients such as Ms M. and Ms K. who are candidates for genetic testing, the decision making lesson is now over. In ninety minutes they have been taught about genes, potential genetic disorders, alleged personal disease risks, and testing and monitoring options. In so doing, the geneticist has given her clients the input that, from her perspective, they need to make a "self-determined decision". How they are to make this decision the geneticist has not explained. As she quite obviously sees no reason not to undergo testing, she emphasizes the principle of self-determination and leaves the pros and cons to her clients' imagination.

The genetic decision making lesson is quite different when a pregnant woman comes in for counselling. Since amniocentesis carries with it the risk of an intervention and can call a pregnancy into question, the geneticists teach their client at length about the pros and cons of their options. They go into detail about the various chances and risks of the options "test" and "no test", and explain the reservations for each one. They instruct the pregnant woman not only about genes and risks but also about risk-related decision making. To arrive at an "informed choice", the pregnant woman learns, she must make a risk assessment – and consider which risks she wants to take.

3.3.4.1. Amniocentesis: An arbitrary test?

Ms A. comes with her husband to the genetic counselling centre at a university clinic because she is 39 years old. Her physician recommended that she undergo amniocentesis testing. At the clinic women receive genetic counselling before they make an appointment for amniocentesis. Such routine counselling is fairly quick: after a 30-minute session the counsellor has prepared the expecting parents for their decision making task. Ms A. is run through a similar counselling programme as Ms G. (see section 3.2.4), though in less detail. The counsellor generates a risk profile of the expecting mother and then teaches her about possible chromosome anomalies, about her so-called age risk, and finally about the pros and cons of amniocentesis.

The counsellor is a gynaecologist with many years of counselling experience. He heads the gynaecology department at the university clinic and performs the amniocentesis himself. Even though amniocentesis testing is his "daily bread", by no means does he want to urge his clients to undergo testing. When questioned by a pregnant woman in another counselling session he openly admits that he would not want his wife to undergo amniocentesis testing. Ruling out Down's syndrome is not important to them. Yet, he would not advise against it. It is very important for him that his clients make their own informed decision.

Ms A. explains at the beginning of the session why she has come to the genetic counselling centre: "I am 39 years old, that is the <u>problem</u>", she says. The correlation between the mother's age and the frequency of chromosomal trisomies has made her aware that she has a "problem" — a problem that she hopes can be solved through counselling and testing. The counsellor goes through his routine counselling programme. First he asks the couple about diseases and any abnormalities among their families and relatives. When he can no longer uncover any additional risks, he begins talking about amniocentesis testing. He explains that it can routinely detect chromosome anomalies. After enquiring about the age of the prospective parents, he calculates a risk figure:

> C: Um, then with this age combination we would have a risk of approximately 2–2.5 percent, (--), uh, let's say, 2 percent for chromosomal disorders.

That he includes the age of the prospective father is completely outdated. Apart from this, he assigns a risk to Ms A. and her unborn child. Then in the following sentence he makes it clear that he cannot say what this risk means for Ms A. Whereas Ms M.'s geneticist delivers the paradoxical assessment that a risk of 80–85 percent does "not say anything personal" but is menacingly high, this counsellor withdraws from the whole affair — he asks his client to assess the risk herself:

> C: First of all this is only a number. Two out of a hundred, one could consider that to be high or low. For good reason it is a fully personal assessment. Um, at any rate, with this scope the test can offer you (-) certainty.

In other words, with a scope of 2-2.5 percent Ms A. can find certainty. This gain in certainty, according to the counsellor, is the advantage of amniocentesis. But: it also has its price.

> C: The disadvantage of amniocentesis, which I have not yet stated, is that in rare cases it can trigger a miscarriage.
> W: That's what I'm afraid of.
> C: This does not happen so often that it needs to be dramatized, but you need to know it. If we perform it 200 times, it happens once, and we don't know with whom.

As with most of his colleagues, the counsellor is silent about the other consequences of amniocentesis testing (Braun, 2006, A 2612; Samerski, 2002). He does not explain to the prospective parents that they must reckon with unexpected and unclear results, nor does he explain what will happen if the test comes back positive. If the test does not deliver the desired gain in certainty but rather a positive test result, then he would force couple A. to make a dramatic decision: they must consider whether Ms A., despite the chromosomal anomaly, will still carry the pregnancy to term, or whether she will have it aborted. Geneticists do not like addressing this topic — after all, they are sitting across from a pregnant woman who most likely is excited about the child she is expecting. Moreover, such an abortion is not just a simple suction procedure but rather a hormonally triggered birth in the fifth month. Apparently, however, this is not anything that should be made clear to a pregnant woman when enabling her to make an "informed decision". In many counselling sessions the counsellors as well as the clients simply ignore the question of what follows in the event of a positive result. During the counselling of couple A. the topic of abortion was not mentioned once.

After the gynaecologist explains to the couple about the risk of chromosomal aberrations with the offspring as well as the chances and risks of amniocentesis, he makes it clear that he can no longer help them. He emphatically prompts the couple to make their own decision. As an expert he has no advice:

> C: There is no, uh, there is no medical standpoint whereby you would need to be told, now you must do this or that. It's not that simple.

As a matter of fact, amniocentesis testing cannot be medically indicated (Schmidtke, 1995). Even if the counsellor later justifies the intervention to the health insurance company with the so-called age indication, the term "indication" is misleading in this context. For one, because a prenatal chromosome test can provide no grounds for treatment, but only grounds for an abortion. Findings for which prenatal treatment, for example hormone therapy, is possible are an exception. And physicians, unless they want to become a eugenicist or a psychiatrist, cannot advise anyone to undergo such an abortion. The term "indication" is also misleading because there is nothing wrong with women like Ms A. or Ms G.: they are not candidates for testing because they have any complaints or symptoms or because any suspicion exists, but solely because of their statistical classification—that is, because certain trait variables have placed them in a risk group. A physician cannot determine an indication based on a probability, "which does not say anything personally", to quote Ms M.'s geneticist. From a statistical standpoint, says the human geneticist Schmidtke, it is impossible to rule out a genetic risk for any woman. Therefore he wants to offer prenatal testing to all women. And teach them how to make "informed decisions".

3.3.4.2. Prenatal decision making and economic rationality

Amniocentesis testing is not a medically recommended diagnostic or treatment measure; it is a risk management option. An option for which there is no medical "ought", and for which women should therefore make their own decisions. The geneticist has just taught Ms A. and her husband what they need to know in order to make an informed and knowledgeable decision. They learned that they have two options, namely, "test" and "no test". Both options, according to the counsellor, bear risks. Their task is to choose one of the risk-laden options. They must consider which of the risks they want to "clarify" and which they want to "accept". After he attests a two percent risk of a chromosome aberration to the mother, he explains what question they now face:

> C: [...] the question is, to ask is this risk so severe that, um, that you want it clarified at the price of accepting the one-half percent risk of miscarriage.

Since there is no "medical standpoint" for the test, Ms A. cannot do anything wrong. The counsellor convinces his client that this risk assessment is something personal. Just as Ms M.'s counsellor repeatedly emphasized that the decision was very personal and that she wants to find the "right path" together with Ms M., this counsellor also stresses that this is about searching for the personally "appropriate" decision:[60]

> C: It is about doing what is appropriate, right? What is suitable for your current situation? And there is no right or wrong.
> W: Hm=hm. Hm=hm.

What the counsellor is demanding of couple A. is paradoxical. On the one hand, he forces them to make their own personal decision on the grounds that there are no medical or scientific criteria. Then, on the other hand, he teaches them to base their decision on statistical constructs. Ms A. learns that as a pregnant woman she must balance risks and accept risks. With the option "no test" she is accepting an elevated risk of bearing a child with a chromosome mutation. With the amniocentesis testing option she is accepting the risk of provoking a miscarriage. To find the right path, the pregnant woman is supposed to weigh these

---

[60] Not only here in this clinic or in Germany, but everywhere in Europe and the USA, genetic counsellors represent this decision as something "personal": see, for example, Bosk (1992), Pilnick (2002), Rapp (1999).

risks. What this means, exactly, the geneticist explains a few sentences later in response to a question from the prospective father:

> M: Now, again, to put this (--) mathematically: The risk of a Mongoloid child is 100:2, right?
> C: Yes, that is, 2 percent, 2 out of 100.
> M: Yes, the risk of a miscarriage is 200:1.
> C: Yes
> M: Okay.
> C: If we now, if we can imagine that 200 women were in their, or 200 families were in your situation, and we performed an amniocentesis on all these women, then we would expect, um, out of these 200 amniocenteses to find four chromosome anomalies, namely, two out of each 100. But we would also have to accept that out of these 200 amniocenteses we would trigger one miscarriage. We don't know with whom.
> W. ((gasps)). mh, yes.

The counsellor first discusses that couple A. should decide what is "appropriate" for their situation. His words suggest that the decision is something personal, about finding their own path. What he now teaches them to prepare them for their decision making, however, is not at all personal. He teaches the prospective parents to make a highly abstract risk assessment: they are to offset four chromosome anomalies out of two hundred pregnant women against one miscarriage out of two hundred pregnant women.

How can a counsellor foist such an abstract cost-benefit analysis on his pregnant client as a way to increase responsibility and autonomy? The "enthronization" of "rational choice", as the German sociologist Uwe Schimank writes, "is a feature of modernity" (Schimank, 2005). Today's meaning of rationality and choice, however, was shaped in the laboratories of the decision scientists in the middle of the twentieth century and has not much in common anymore with the meaning of these terms in the time of Max Weber. Originally, the colloquial term "decision" stems from the activity of separating, a meaning which is still included in the juridical term of a judicial "decision". The American Heritage Dictionary of the English Language (AHDEL, 1992) lists as a first denotation of *to decide*, "the passing of judgment on an issue under consideration", and as a second "the act of reaching a conclusion or making up one's mind". As a synonym of *decision* it does not mention *choice*, but "conclusion" and "determination". The Oxford English Dictionary (OED), too, explains *to decide* as "to determine (a question, controversy, or cause) by giving the victory to one side or the other; to bring to a settlement, settle, resolve (a matter in dispute, doubt, or suspense)". A further meaning emphasizes the judgment in the context of settling a dispute: "To settle a question in dispute, to pronounce a final judgment" (OED, s.v. "decide"). In the second half of the twentieth century the meaning of decision was fundamentally transformed. Next to the juridical sense of the term, reference books of the 1970s introduce a completely new notion of decision. They list numerous neologisms such as "decision logic", "decision process", "decision goals", "decision rules", "decision problems". But, not only was there constituted a new knowledge field of "decision" with new objects and technical terms, but also the meaning of the colloquial word "to decide" was explained in a new way: since the 1970s, to decide is not synonymous with "resolve" or "determine", but with "choice between options".[61]

It was decision theory that fabricated this new notion of decision, with its corresponding neologisms. After World War II, behaviourism, utility theory, systems analysis, probability theory, and statistics have constituted a new interdisciplinary research field that aims at calculating and optimizing choices. In decision theory, "decision" designates a mechanical act of choice, namely the selection of an option as a consequence of an algorithm. It presumes that decisions rely on information processing, likewise in humans, machines, or other systems. Their basic point of view, Nobel Prize laureate Herbert Simon wrote in 1972, was "that the programmed computer and human prob-

---

[61] For the transformations in the German language, see Samerski (2002).

lem solver are both species belonging to the genus IPS [information processing systems]" (Newell and Simon, 1972, p. 870). As historian of science Lorraine Daston has poignantly put it: mindful judgment is thus transformed into algorithmic rationality (Erikson *et al.*, 2013). A central topic of decision theory is decision making under risk. Nothing in the world is certain, decision scientists declare, and therefore we are forced to calculate and evaluate risks and probabilities. No matter if it's health, car driving, or financial investments: realizing and weighing risks, they say, is the epitome of rationality. This is why cognitive psychologist Gerd Gigerenzer has founded the Harding Center for Risk Literacy and advocates a risk literacy curriculum at schools. As head of the German Max Planck Institute for Human Development, he propagates the critically applied probability calculus as the acme of enlightened reasoning (Gigerenzer, 2003; 2014). Only those who are "risk savvy", he claims, can be make the right choices.

The geneticist introduces the prospective parents to decision making according to the tenets and rules of this algorithmic rationality. In order to make a decision about the pregnancy, they are expected to assess statistical risks and balance them. They are to weight their personal preferences to the probabilities of events and then calculate the resulting risks. This is exactly the decision technology recommended by decision theorists to maximize "expected utility". Albeit controversially discussed and convincingly challenged, it is the fundamental model of decision making in economics up to the present day.[62] The anticipated, statistically calculated costs are offset against the anticipated, statistically calculated benefits. If a health economist wants to verify the cost effectiveness of the measure "amniocentesis", she would use such a method. She would

---

[62] For criticism of rational choice theory, see, among others, Amartya Sen's *Rational Fools* (1977). S.M. Amadae (2003) and Philip Mirowski (2002) have critically examined the advancement of the decision sciences and their influence on economics, political science, and society.

balance the statistically ascertained costs and benefits.[63] To do so, she would compare the incidence of undesirable side effects, including induced miscarriages, with the incidence of desired outcomes, namely, the detection of chromosome aberrations. Depending upon the enquiry, she may then recommend changing certain parameters of the measure to optimize the relation between costs and benefits: thus, for instance, changing the test population could lead to an improvement in the health economic benefits. The first trimester screen,[64] for example, has exactly this function: as two human geneticists claim, it effectively reduces the rate of invasive diagnostics and the consequent intervention-related abortions with a simultaneous increase in the rate of detection (Eiben and Glaubitz, 2005). A risk balance therefore serves to weigh statistically measured and assessable overall results. But for the pregnant woman sitting in front of the geneticist, such managerial decision making is not only insentient, but irrational. Overall statistical results are irrelevant to her — she cares about the child whom she will be bringing into the world. Hence, most women are desperate to wring any meaning from the risk (Rothman, 1986; Samerski, 2002; Schwennesen, Koch and Svendsen, 2009). In the genetic choice training Ms A. is asked to make as her own a rationality

---

[63] The question about benefits is here naturally inhuman. Time and again economic studies are conducted that offset the financial costs of amniocentesis against the savings made when a child with Down's syndrome is not born into the world. They conclude that amniocentesis is beneficial to the economy. The price, however, which particularly women, and ultimately all members of society, pay when people who do not fulfil our ideas of the norm are prenatally selected is naturally not taken into account by the economists.

[64] The first trimester test is a screening technology that results in a statistical risk for delivering a child with Down's syndrome. It is performed between the 11th and 14th week of pregnancy. Sonographic markers such as the nuchal translucency, maternal age, and maternal serum levels are fed into a computer program that calculates a probability. This number indicates frequency of Down's syndrome in the population of women with this risk profile. Thus, the first trimester screen is not a diagnostic test.

which principally abstracts from incarnate persons and their concrete circumstances. She learns to incorporate a "statistical gaze" (Rieger, 2010) and to select between options on the basis of calculated values. To apply such an economic calculation to the coming of a child is not only inhuman but also absurd.[65]

By inviting them to weigh risks and make a managerial decision, the counsellor is not instructing his clients in a rationality outside of his field; rather, he is introducing them to the logic underlying his service. The entire field of prenatal diagnostics is built on balancing risks. When, for example, a pregnant woman is attested a "high risk" for a child with Down's syndrome, then this—seemingly medical—attest is based on a cost-benefit assessment. Whether an age risk or as a first trimester screen classified as "positive": a risk is classified as "high" and is thus considered as grounds for testing when it is greater than 1:250–300. This cut-off point has no medical grounds; rather, it is based on a risk balance. The probability of causing a miscarriage with amniocentesis is given at around

---

[65] I deliberately avoided interviewing counselling clients after their sessions. How they experienced their counselling sessions I was only able to read from their direct reactions. However, how confused, perplexed, unmoved, or annoyed the clients of genetic counselling can be, and what burden the "informed decision" places on them, comes up in numerous interview studies. The genetic-medical jargon itself often intimidates clients (Chapple, Champion and May, 1997); not long after the session clients are no longer able to recall the main information or have interpreted it their own way (Friedrich, Henze and Stemann-Acheampong, 1998; Lippman-Hand and Fraser, 1979a,b,c; Lock, 2009); they have been promised an authoritative knowledge, which, however, does not offer certainty but only widens the spectrum of scary future scenarios (Sarangi, 2002; Zuuren, Schie and Baaren, 1997); and finally, many counselling clients experience the decision about the test not at all as a free choice but as an obligation (Hallowell, 1999; Schwennesen, Koch and Svendsen, 2009); or they do feel informed, but helpless and stuck (Lippman-Hand and Fraser, 1979a,b); or they submit to a test fully unaware of what may happen (Friedrich, Henze and Stemann-Acheampong, 1998). These interview studies impressively show what the call for an "informed decision" based on genetic and statistical constructs demands of counselling clients.

1:200. So that—from a statistical perspective, or rather, in the long-term—no more miscarriages are caused than trisomies are detected, amniocentesis is mainly offered to the female population for whom the risk of bearing a child with Down's syndrome is higher than the risk of a miscarriage. Pregnant women attested with a trisomy probability of less than 1:250 are therefore not among the target group of test candidates. They are considered "inconspicuous" and receive amniocentesis testing only when they insist. Pregnant women with a trisomy probability greater than 1:250, on the other hand, belong to the target group of test candidates. They are considered "conspicuous" and in discussions with their physicians and in genetic counselling sessions they are called upon to make an informed choice about undergoing chromosome testing.

In another counselling session a geneticist explicitly conducts a cost-benefit assessment based on the age risk. While she explains the amniocentesis test, she shows her pregnant client a table which depicts the probability of trisomy 21 linked with age. She comments:

> C: There is, you know, a bit of a miscarriage risk. It's low, but it certainly isn't zero. Otherwise it would... otherwise here ((she points at the table to the age under 35)) more healthy children would be lost than that something would be detected. Starting here ((she points at the table to the age of 35)) it begins to stand in relation to what could be detected.

The so-called age risk is nothing other than an instrument for generating a test population for which chromosome testing is statistically worthwhile. It is a marker used to generate a target group for further testing. If every pregnant woman underwent amniocentesis testing, then the overall balance would show the loss of more "healthy children" than the detection of chromosome aberrations. If, however, a population of pregnant women is selected in which the probability of positive test results is greater than the probability of triggering a miscarriage, then the overall balance will be positive. The so-called age risk as well as the first trimester screen or the triple test serve as a screen to generate test populations. They ascertain biological and sonographic markers by means of which women are ultimately assigned a risk profile—a risk profile that makes them a potential test candidate or not.

The geneticist has now asked Ms A., based on this cost-benefit logic, to make a decision about her pregnancy. She is expected to weigh "2 out of 100" and "1 out of 200" as well as "miscarriage" and "gain certainty". The naked probabilities do not suffice for the decision making calculations: they only define the framework in which the decision is to be made. Ms A. must add her own personal assessment to the risk figures, her personal preferences in order to maximize utility, as decision theorists would say. She is to amalgamate her desires, fears, and hopes with the risks in order to arrive at her own purportedly rational decision. The geneticist expressly asks her to personally assess the risks and to conduct an individual cost-benefit assessment:

> C: If at the end of your assessment the <u>advantage</u> of an amniocentesis is greater than the disadvantage of the miscarriage risk, (-) if that is the case, then testing is certainly appropriate. But if not, and you fear the miscarriage risk of the amniocentesis (-) more than you value the advantage, then you'd better not do it.

Ms A. is to combine calculated risks with her personal preferences and then conduct a risk assessment. Thus, the geneticist is imposing on the expecting mother to act according to the dictates of an economic rationality.

3.4. The Decision Trap

The mindset and way of thinking demanded by "informed choices" is made very clear with the example of prenatal decision making lessons. Facing a pregnant woman, geneticists painstakingly spell out the decision required by risk attestations and test options. Emphatically and step by step they teach their clients an economic style of thinking, a managerial decision making strategy. Yet the example of pregnant women shows not only what mode of thinking underlies "informed choice" but also what price it has. Rothman trenchantly describes the real consequences of an abstract rationale for a pregnant woman who "enters into a rational seeking of information and choices and finds herself trapped in a nightmare" (Rothman, 1986, p. 181). The demand for decision making based on options and risks not only challenges women's trust in themselves, their intuition, and their assurance, but also much more: their coming child. By educating her about the risks, testing options, and alternatives the genetic counsellor persuades his client that the outcome of her pregnancy is the object of her decision making. She is being asked to consider whether she wants to bear the unborn child in light of its current risk profile or whether she will make its birth dependent on further tests. She is to become a *manager* of the person-to-be whom she carries under her heart.

Such managerial decision making requires a form of perception and thinking that until recently has had nothing to do with pregnancy and becoming a mother. The prenatal risk assessments, that women today are routinely asked to do, destroy the mindset that only a few generations ago was the symbol of pregnancy: good hope. Until prenatal diagnostic testing was introduced in the 1970s, women who were expecting a child did not have to make any decisions.[66] They were in "other circumstances" and tried to avoid bad thoughts, unstilled desires, any terrifying sight, or other harmful influences. But no matter what they did or did not do—the outcome of their pregnancy remained uncertain (Duden, 2002b). Before biology and medicine objectified the unborn child as a fetus in the nineteenth century, pregnancy was not a physiological and objectifiable state (Samerski, 2014). "Pregnant" was the expression for the particular mindset of the woman toward her unborn child, her somatic *hexis*, or physical disposition[67] (Duden, 2000).

---

[66] I am talking here expressly about women who are expecting a child and not about those who shudder at the thought and have no desire to be of "good hope".

[67] Barbara Duden does not talk about the "physical" but the "somatic", because she is referring not to the biologically defined "body" of science nor the abstract "body" in academic discourse, but rather the experi-

Women were, as Germans so eloquently say, "of good hope". The reality of what was already latently present in them would only be revealed at birth. Birth marked the threshold between two heterogeneous spheres: inside and outside, invisible and visible, beyond and here and now. Birth, as philosopher Hannah Arendt says, is beginning and surprise. Humans again and again marvel at the newborn, the completely new and singular: "The new always happens against the overwhelming odds of statistical laws and their probability, which for all practical, everyday purposes amounts to certainty; the new therefore always appears in the guise of a miracle" (Arendt, 1958, p. 178).

Ultrasounds, prenatal tests, and risk-related pregnancy care make this particular female mindset practically impossible. Biology and medicine have reframed pregnancy into a biological developmental process that must be monitored and managed by experts. The centre of medical attention is no longer the pregnant woman, but instead a technogenetic construct: the fetus and his norm-appropriate development (Samerski, 2014). Sonographic monitoring and laboratory tests monitor fetal growth for risk factors and potential deviations. The task of the pregnant woman is now to participate in monitoring and managing her womb. It is considered to be the obligation of a responsible mother to manage the unborn child's risk profile. She is expected to reduce fetal developmental risks and create optimal conditions for its growth, for instance by taking vitamins, not consuming cheese, and undergoing ultrasounds and chromosome tests (Samerski, 2009).[68]

---

enced body. The Greek work *hexis* can be best translated as "second nature", habit, or disposition.

[68] Oakley (1984), among others, has examined in detail the history of the medicalization of pregnancy and birth. William Arney (1982) demonstrates how medicine over the course of the twentieth century increasingly understood pregnancy and birth as an enterprise in need of regulation and surveillance and in which the woman must be actively involved. Lorna Weir (2006) has dedicated a book to the loss of birth as a milestone in human life. She argues that starting in the 1950s the concept

The genetic decision making lesson further reinforces this transformation of the expecting mother into a fetal risk manager. The geneticist obliges the pregnant woman not only to forego cheese and undergo ultrasounds but also saddles her with a completely new obligation: she herself must choose and feel responsible for the risks she takes. No longer can she merely fulfil her obligation by complying with medical guidelines. Folic acid supplementation, regular doctor visits, and ultrasound screenings no longer suffice. She can no longer indulge in the exonerating illusion that she will have done everything right if she just follows the doctor's orders: she herself has now been made responsible for the management of the fetal risk profile — and thus also for the outcome of her pregnancy. Self-help literature, gynaecologists, and geneticists drum into her that she herself must know how to deal with the risk profile she is carrying inside her: which tests she undergoes, how she assesses the risks, and what conclusions she draws from them.

The pregnant woman is thus practically compelled to make risk-related choice about pregnancy — no matter whether she is to decide about a first trimester screen or amniocentesis and whether she ultimately says "yes" or "no". Even when the test is treated as a routine part of a normal pregnancy trajectory, pregnant women are explicitly asked to make a choice; they experience "no choice but to choose" (Schwennesen, Koch and Svendsen, 2009, p. 196). Ms A.'s genetic counselling reveals this compulsion to managerial decision making about the unborn child. As a pregnant woman, says the counsellor, Ms A. has two options: either she gives birth to her child without any prior testing and thus accepts a two percent risk that the child may have a trisomy; or she lets herself be tested and accepts the risk of causing a miscarriage as well as the possibility of having to decide whether to continue her pregnancy in the event of a

---

of "perinatal mortality" opened up new terrain for medical research, legal regulations, medical interventions, and state and professional disciplining: intra- and extra-uterine development.

positive finding.[69] The counsellor has presented both possibilities, test or no test, as options that require a decision. In either case the pregnant woman herself must bear the responsibility for what might happen afterwards.

If Ms A. cannot elude the logic of the lesson in genetics, she subsequently sits in a trap: in the decision trap. There is nothing else she can do but choose between risk-related options. She must abandon her good hope and resolve to weigh her options.[70] This is the symbolic power of the counselling and educational procedures surrounding modern pregnancy: women are persuaded that calculated decision making about their unborn child is among the obligations of a responsible pregnant woman. Once they land in the decision trap there is only "informed choice". The prenatal decision making lesson turns even a refusal into a predetermined and risk-laden option. The fundamental difference between a "yes" and a "no" to prenatal screening and "tentative pregnancy" (Rothman, 1986) vanishes. A refusal becomes the acceptance of an analogous

---

[69] According to estimates, far more than 90 percent of pregnant women whose child is prenatally placed in the diagnostic class "Down's syndrome" terminate their pregnancy. Even findings with only slight or perhaps not any consequences convey to prospective parents the message "The child is not normal", and cast a shadow on the pregnancy (Rothman, 1986). The prospective child, which they have not yet been able to see and take into their arms, becomes a modern changeling, a menacing foreign body: "I was overcome with the horrible feeling of having a 'monster' in my belly. My first impulse was a strong desire to get rid of it as quickly as possible", said a pregnant woman after a positive ultrasound finding (Schirmer, 2009, p. 122).

[70] This obligation to make a decision and bear the responsibility is likely the overriding symbolic function of the prenatal testing industry. The paradox is that freedom is becoming more constrained on legal, actuarial, and social grounds, and for everyone involved, while pregnant women are increasingly called upon to make their "own decisions". Prenatal diagnostic testing permits very little decision making leeway for everyone involved. It is becoming more difficult to decide not to undergo certain prenatal tests, as psychosocial counsellors observe (Braun, 2006, p. 2). Therefore the call to make one's "own decision" should be interpreted as a ritual with a symbolic function.

option. Even if she simply wants to be pregnant and to look forward to her child, the pregnant woman must expressly decide to continue her pregnancy and to accept the predicted risks.[71] Thus, the decision trap renders impossible what motherhood meant for my mother's generation: bear her coming child without any ifs or buts. Today when a child is born, the pregnant woman is no longer of good hope but has made an informed decision. Not a decision to start a family or to carry the pregnancy to term despite an absconding father, no — she had to make a completely different choice: a calculated decision to give birth to this risk profile and to feel responsible for it.[72]

---

[71] "The risk figure was explained to me in a very professional manner, but I was not able to make sense of it...", complained a pregnant woman. Nevertheless she was pushed in the genetic counselling session to make a decision and to feel responsible for it: "They emphasized over and over again, that is, well, it is your choice... So I had to take responsibility" (Schwennesen, Koch and Svendsen, 2009, p. 202). "Pregnant women", concludes Anne Balsamo, "are both disempowered and held responsible at the same time" (Balsamo, 1996, p. 110).

[72] Naturally, I do not want to equate here what is demanded of pregnant women with what they actually do later. Interview studies show how difficult this decision making is for women. Most of them probably do not make an informed choice or they do it in an outright schizoid, that is, split, state of mind. Many studies point out that a fundamental discrepancy exists between theoretical decision making on the basis of abstract deliberations and the existential dismay that becomes acute when the test is looming and thus also the possibility of a conspicuous finding (Friedrich, Henze and Stemann-Acheampong, 1998, p. 106); see here also Rothman (1986). A majority of the women interviewed, observed the Göttingen medical sociologists Hannes Friedrich, Karl-Heinz Henze and Susanne Stemann-Acheampong, showed invariably characteristic confusion, contradictions, and gaps in memory, and occasionally downright amnesia when it came to the basic facts and processes (Friedrich, Henze and Stemann-Acheampong, 1998, p. 118). By no means could the confusions be attributed to inadequate instruction. The authors thus conclude that obviously it is hardly possible to make the decision for prenatal diagnostic testing in accordance with biographically significant experiences and with one's immediate psychological and physical experience (Friedrich, Henze and Stemann-Acheampong, 1998, p. 106).

Chapter Four

Conclusion
Disempowering Autonomy

4.1. The Tyranny of Choice

What *can* be decided, must also *be* decided. Life today is no longer governed by tradition, nature, and fate—according to a sociological truism—rather it has become dependent on autonomous decisions, even biology itself (Beck and Beck-Gernsheim, 1994; Beck-Gernsheim, 1995). Accordingly, not only do we choose profession, partner, and lifestyle, but also what has hitherto been given as "body" or "nature" and seemed unavailable. Science and technology can control and shape human life in its molecular bases, while deferring the decision making to us. This is the tenor of numerous social scientific analyses that examine the new compulsory choices of our time. With this in mind, they also declare informed decision making about genetic testing to be an inevitable necessity, even human destiny. Even authors who want to reveal the specific power of biopolitics spread this myth of technological feasibility.

Modern biopolitics is characterized, says Nikolas Rose for example, in that it can shape and guide people as biological creatures: "Vital politics... is concerned with our growing capacities to control, manage, engineer and reshape, and modulate the very vital capacities of human beings as living creatures" (Rose, 2007, p. 3). What Rose calls "autonomization" and "responsibilization" in health care, in other words imposing "self-determination" and "responsibility", appears then to be the inevitable consequence of these new feasibilities. In the face of new genetic tests, sociologist Thomas Lemke justifies the new decision making constraints by claiming there is no way to avoid genetic self-determination: even saying "no" to testing is a form of selection, namely the selection of a "natural genetic make-up". "Whether we like it or not, even the seemingly 'non-eugenic' decision against genetic diagnostics and selective abortion has a eugenic quality, since it is based on a (normative) decision: the decision that it is better not to decide. The choice of a 'natural' genetic make-up for an individual is only one option and one 'selection' among others, in any case it is an option — neither fate nor unchangeable" (Lemke, 2005, p. 198). Here Lemke concedes to genetics the full interpretive authority over reality: he reframes saying "yes" to a coming child into a decision in favour of a genetic configuration. Thus according to these authors there is no longer a standpoint beyond that of "genes" and "informed decision making". Saying "no" is thus one of the predetermined options within this rationality against which this "no" is actually directed. We choose our destiny not only through testing, according to Lemke, but also choosing has become our destiny.

The objective of genetic education is to prepare citizens for this new self-determination. The contradictions and aporias they face as a result are very clear in the counselling sessions of Ms M., Ms K., Ms G., and Ms A. On closer inspection the promise of "knowledge", "decision making", and "feasibilities" turns out to be a trap. The counselling clients are convinced that they must decide and take risks without being able to influence what actually affects them. The expertise of geneticists, presented as meaningful "knowledge", and the services offered are based on statistical artefacts: on risk profiles and possible events in populations. These possible events — whether the development of cancer or disabled children — are anticipated by geneticists as calculable risks. Based on a client's statistical traits the geneticist determines how high these risks are by placing the client in a relevant population and then ascribing to this client the probabilities of this population as alleged personal risks. Ms M., Ms K., and Ms A. can influence this risk profile in their decision making. If they choose one of the predetermined options, they change a trait, are subsequently placed in another population, and thereby modify how high their attested risks are. If, for example, Ms K. chooses the option of genetic testing, then—by incorporating a genetic trait—the incidence rate of developing cancer changes. She is placed in a new statistical collective, and is ascribed their probabilistic characteristics as a new risk. If Ms A. chooses the option "amniocentesis", she minimizes the risk for microscopically recognizable chromosome mutations, but thereby calls into play the risk of an induced miscarriage.[1] This correlation between traits on the one hand and events in populations on the other hand suggests that what may happen to someone in the future can be manipulated in the present. By means of statistics and the theory of probability the counsellor constructs an anticipated future that seems calculable and accessible.[2]

What actually happens to Ms K. and Ms A., however, is another story. Even if they dutifully manage their risk profile, it will not influence what is meaningful to them. What will

---

[1] The decisions made by clients influence not only their risk profile but also the general basis for risk calculation, namely the corresponding statistical data and often also the algorithms used. The first trimester test, for example, was long designed as a big experiment: physicians who wanted to perform the test had to submit their results to Kyprus Nikolaides' London Prenatal Centre where the corresponding calculation software with the so-called Nikolaides algorithm had been developed and was optimized with the incoming data.

[2] Also Schwennesen, Koch and Svendsen (2009) conclude in their empirical study on the first trimester test in Denmark that the demand for an "informed decision" burdens women with responsibility for the future. "With the principle of informed choice, and the ideal of the pregnant woman as an autonomous individual, the woman and her partner are constituted as responsible for the choice they are making and thereby also for the future which is created through their decision making" (Schwennesen, Koch and Svendsen, 2009, p. 202). Since the authors recognize the gulf between the epidemiological risks and the concrete pregnant women, but do not perceive the epistemological difference, they overlook the gulf between predicted risks on the one hand and what later actually happens and is experienced.

eventually happen to them cannot be determined by risk management. Whether Ms K. will one day have a lump in her breast, at what age, and what course the disease will then take, remains unknown. And whether or not Ms A. will bear a healthy child that will lead a normal life cannot be anticipated by anyone. Nevertheless, the counselling session has placed both women in a dilemma. The geneticists have convinced them that no matter what they must make a decision and accept the ensuing risks. Now they are caught in the decision trap.

This obligation to make a decision and bear the responsibility is the overriding symbolic function of the professional facilitation of "informed choice". Oftentimes, the imperative of "knowledge" and "prevention" permits very little decision making leeway for everyone involved, but nevertheless "choice" and "autonomy" are conjured up. Hence, the call to choose should be interpreted as a ritual with a symbolic function. As the counselling sessions have shown, the obligation to make an "informed choice" turns clients into *decision makers* where they cannot do anything and declares them responsible where they are powerless. Now they can be blamed for whatever happens. If Ms K. develops breast cancer, the blame lies with her risky genes that she did not manage carefully enough. If Ms A. bears a disabled child, it was her choice; after all, she decided in favour of giving birth to this risk profile. Genetic education produces decision makers that are at the same time disempowered and held responsible.

4.2. Autonomous Decision Making as Social Technology

Back in 1981 the French social scientist Robert Castel described the management of risk profiles as a new form of social technology based no longer on people but on statistical constructs. People are "labelled", as Castel calls it, with risk profiles, and assigned certain "paths": "We stand before the prospect of an automated management of populations based on differential profiles", as he summarized the new form of a risk-related social technology three decades ago (Castel, 1983). "Automated management", however, has not materialized. Guided self-determination is a much more effective administrative regime than top-down government: citizens are not degraded to passive objects of management, but are mobilized to engage in the risk management of their own affairs.[3] They are not prescribed a certain path but must make a choice. Geneticists assign them risk profiles, offer them various paths as options, and demand that they assume the responsibility for everything else.

When Immanuel Kant coined the term "self-determination" around 250 years ago, he could not have imagined its perversion in genetic education. The expert-directed self-management of the twenty-first century no longer has anything to do with the capacity "to make use of his understanding without direction from another".[4] Today, anyone who wants to be regarded as mature must have their decision making guided by professionals. Today, anyone who wants to be regarded as self-determined must be told by experts what this "self" actually is — and which options are being offered for its "determination". What citizens learn, then, is how to treat themselves as a statistical artefact. In the name of self-determination they are being called upon to deny their subjectivity, even their person. Habermas's terrifying vision of the genetic instrumentalization

---

[3] How risk serves as the basis of and means for new forms of social discipline and social control has meanwhile been widely investigated, see, among others, Dean (1998), Ewald (1991; 1993), Weir (1996). "It is 'an integral part of the various forms of calculating rationality that can control the behavior of individuals, groups, and populations'" (Dean, 1998, p. 131). If the affected persons are denied the capacity of responsibility and self-determination, then the risk technologies have a definite link to authoritarian forms of social discipline, see Kaufert and O'Neil (1993), Weir (2006).

[4] "Enlightenment is man's release from his self-incurred tutelage. Tutelage is man's inability to make use of his understanding without direction from another. Self-incurred is this tutelage when its cause lies not in lack of reason but in lack of resolution and courage to use it without direction from another. Sapere aude! 'Have courage to use your own reason!' — that is the motto of enlightenment" (Kant, 1784/1997).

of human beings has already become reality (Habermas, 2001):[5] not the genetically engineered human being, as Habermas believes, but its intellectual and ideological foundations that blur anthropologically deep-seated categorical distinctions between subjective and objective, the grown and the made (Habermas, 2001, p. 112). It is not genetic manipulation that transforms the unique human being into something calculable and feasible, but the obligation to make an informed choice.

Which abysses open is manifest especially when a pregnant woman is called upon to make an informed decision. Genetic education calls into question not only her self-perception and her sense of the present but also her attitude toward the becoming "you". The lessons on genes and risks ask her not only to deny her own person but also the uniqueness of her unborn child. By advising his client to engage in managerial decision making about her pregnancy, the geneticist is asking her to dehumanize what she carries under her heart. For the "informed decision", to which he is guiding her, she is to anticipate the "you" growing inside her as a risk profile, as a faceless case. And then to consider whether she wants to accept it or not.

4.3. Conclusion: Now What?

Whenever I give a lecture or an interview, I am often asked: "Now what? Should we let physicians make the decisions again?!" I have no magic bullet to offer. On no account is it my intent to push the envelope of obligations and decision making powers or to join the chorus of those who demand more professional counselling—patient-centred, holistic, or psychosocial in addition to medical. My concern lies elsewhere: I am

---

[5] He voices concern that, in the future via manipulation of the genome, we will encounter not only born persons but also "created" ones (Habermas, 2001, p. 112). By means of their status created persons would undermine the foundations of morality, even the possibility of morality. A genetic instrumentalization of human beings would create entirely new forms of paternalism and dependency and call into question the "fundamental self-image" of human beings, according to Habermas.

interested in understanding the present in order to estrange myself from it. I want to open for myself and others the experience that one "can think differently than one might think, and can perceive differently than one sees" (Foucault, 1986, p. 15). For this reason I have tried to take power away from the thought constraints of "gene", "risk", and "informed decision making". I want to show that underlying them is a form of thinking that turns people into calculable constructs, into faceless risk profiles. In doing so, I hope to give people courage: courage to trust in one's heart and mind and not to let oneself be lead into the decision trap—so that your garden tomatoes will not turn into "G.M.O-food" and your neighbour will not mutate into a gene carrier.

Transcription Conventions

after Selting (1998)

C:	Counsellor
W:, M:	Woman, Man
(-), (--), (---)	short, middle length, and long break
(2.0)	Break of approximately 2 seconds
:, ::, :::	Elongation (of a vowel)
=	Immediate connection without break
<u>not</u>	emphasized
((laughs))	Nonverbal events

Bibliography

Adelswärd, Viveka, and Sachs, Lisbeth (1996) The meaning of 6.8: Numeracy and normality in health information talks, *Social Science and Medicine*, 43 (8), pp. 1179-1187.

Allen, Garland (1997) The social and economic origins of genetic determinism: A case history of the american eugenics movement, 1900-1940 and its lessons for today, *Genetica*, 99, pp. 77-88.

Amadae, S.M. (2003) *Rationalizing Capitalist Democracy: The Cold War Origins of Rational Choice Liberalism*, Chicago: University of Chicago Press.

American Society for Human Genetics (2004) *Enhancement of K-12 Human Genetics Education: Creating a Cooperative Plan*, Education Summit 9-10 September 2004, Bethesda, Maryland.

Arendt, Hannah (1958) *The Human Condition*, Chicago: University of Chicago Press.

Arendt, Hannah (1963/2007) The conquest of space and the stature of man, *The New Atlantis*, Fall 2007, pp. 43-55.

Armstrong, David (1995) The rise of surveillance medicine, *Sociology of Health and Illness*, 17 (3), pp. 393-404.

Armstrong, David, Michie, Susan, and Marteau, Theresa M. (1998) Revealed identity: A study of the process of genetic counselling, *Social Science and Medicine*, 47 (11), pp. 1653-1658.

Arney, William R. (1982) *Power and the Profession of Obstetrics*, Chicago: University of Chicago Press.

Arney, William R., and Bergen, Bernhard J. (1984) *Medicine and the Management of Living: Taming the Last Great Beast*, Chicago: University of Chicago Press.

Baitsch, Helmut (1970) Das eugenische Konzept – einst und jetzt, in Wendt, G. Gerhard (ed.) *Genetik und Gesellschaft*. *Marburger Forum Philippinum*, Stuttgart: Wissenschaftliche Verlagsgesellschaft, pp. 56–71.

Baker, Catherine (1997) *Your Genes, Your Choices: Exploring the Issues Raised by Genetic Research*, American Association of the Advancement of Science.

Balsamo, Anne (1996) *Public Pregnancies and Cultural Narratives of Surveillance: Technologies of the Gendered Body*, London: Duke University Press, pp. 80–115.

Bauman, Zygmunt (1989) *Modernity and the Holocaust*, Ithaca, NY: Cornell University Press.

Baureithel, Ulrike (2003) Unter Generalverdacht. Zur Renaissance einer Denkfigur, in Nusser, Tanja, and Strowick, Elisabeth (eds.) *Rasterfahndungen. Darstellungstechniken – Normierungsverfahren – Wahrnehmungskonstitution*, Bielefeld: transcript.

Beck, Ulrich (1992) *Risk Society: Towards a New Modernity*, New Delhi: Sage.

Beck, Ulrich, and Beck-Gernsheim, Elisabeth (1994) *Riskante Freiheiten. Individualisierung in modernen Gesellschaften*, Frankfurt am Main: Suhrkamp.

Beck-Gernsheim, Elisabeth (1995) *Welche Gesundheit wollen wir?*, Frankfurt am Main: Suhrkamp.

Begg, Colin B. (2002) On the use of familial aggregation in population-based case probands for calculating penetrance, *Journal of the National Cancer Institute*, 94 (16), pp. 1221–1226.

Bernstein, Peter L. (1996) *Against the Gods: The Remarkable Story of Risk*, New York: John Wileys & Sons.

Berufsverband Medizinische Genetik e.V. (1996) Leitlinien zur Erbringung humangenetischer Leistungen: 1. Leitlinien zur Genetischen Beratung, *Medizinische Genetik*, 8 (3), pp. 1–2.

Berufsverband Medizinische Genetik e.V. (1997) Leitlinien zur zytogenetischen Labordiagnostik, *Medizinische Genetik*, 9, pp. 560–561.

Beurton, Peter J., Falk, Raphael, and Rheinberger, Hans-Jörg (2000) Introduction, in Beurton, Peter J., Falk, Raphael, and Rheinberger, Hans-Jörg (eds.) *The Concept of the Gene in Develop-*

ment and Evolution: Historical and Epistemological Perspectives, Cambridge: Cambridge University Press, pp. ix–xiv.

Beurton, Peter J., Rheinberger, Hans-Jörg, and Falk, Raphael (2000) *The Concept of the Gene in Development and Evolution: Historical and Epistemological Perspectives*, Cambridge: Cambridge University Press.

Bosk, Charles L. (1992) *All God's Mistake: Genetic Counseling in a Pediatric Hospital*, Chicago, London: University of Chicago Press.

Braun, Annegret (2006) Spätabbrüche nach Pränataldiagnostik: Der Wunsch nach dem perfekten Kind, *Deutsches Ärzteblatt*, 103 (40), pp. A–2612–6.

Braun, Kathrin, et al. (2008) Die Sprache der Ethik und die Politik des richtigen Sprechens. Ethikregime in Deutschland, Frankreich und Großbritannien, in Mayntz, Renate, Neidhardt, Friedhelm, Weingart, Peter, and Wengenrodt, Ulrich (eds.) *Wissensproduktion und Wissenstransfer*, Bielefeld, pp. 221–242.

Bröckling, Ulrich, Krasmann, Thomas, and Lemke, Thomas (2010) *Gouvernmentality: Current Issues and Future Challenges*, London, New York: Routledge.

Bröckling, Ulrich, Krasmann, Susanne, and Lemke, Thomas (eds.) (2004) *Glossar der Gegenwart*, Frankfurt am Main: Suhrkamp.

Bundesärztekammer und Deutsche Gesellschaft für Gynäkologie und Geburtshilfe e.V. (2009) Gesetzliche Änderung des Schwangerschaftskonfliktgesetzes. Gemeinsame Stellungnahme der Bundesärztekammer und der Deutschen Gesellschaft für Gynäkologie und Geburtshilfe e.V. für die Anhörung am 16. März 2009, in *Deutscher Bundestag, Ausschuß für Familie Frauen Senioren und Jugend* (Hg.). Ausschussdrucksache 16(13)439f, [Online] http://www.dggg.de/fileadmin/public_docs/Stellungnahmen/0903_stellungn%20baek-dggg_schwangerschaftskonfliktgesetz.pdf [9.1.2015].

Bundeszentrale für politische Bildung (2003) *Veranstaltungsdokumentation (September 2003), Gute Gene – schlechte Gene? Gentechnik, Genforschung und Consumer Genetics*, [Online] http://www1.bpb.de/veranstaltungen/CQQ6WJ,0,Gute_Gene_schlechte_Gene.html [27.3.2010].

Bundeszentrale für politische Bildung (2010) *Die bpb. Leitbild der Bundeszentrale für politische Bildung,* [Online] http://www.bpb. de/die_bpb/WHOLJ2,0,0,Leitbild_der_Bundeszentrale_f%FCr_ politische_Bildung.html [5.1.2010].

Bunton, Robin, and Petersen, Alan (2005) *Genetic Governance: Health, Risk and Ethics in a Biotech Era,* London, New York: Routledge.

Burian, Richard M. (1986) On conceptual change in biology: The case of the gene, in Depew, David J., and Weber, Bruce H. (eds.) *Evolution at a Crossroads: The New Biology and the New Philosophy of Science,* Cambridge, MA: MIT Press, pp. 21–42.

Calderon-Margalit, R., and Paltiel, O. (2004) Prevention of breast cancer in women who carry BRCA1 or BRCA2 mutations: A critical review of the literature, *Int J Cancer,* 112 (3), pp. 357–364.

Caruso, Denise (2007) A challenge to gene theory, a tougher look at biotech, *New York Times,* 1.7.2007.

Castel, Robert (1983) Von der Gefährlichkeit zum Risiko, in Wambach, Manfred Max (ed.) *Der Mensch als Risiko. Zur Logik Von Prävention und Früherkennung,* Frankfurt am Main: Suhrkamp, pp. 51–73.

Castellani, C., et al. (2009) Consensus on the use and interpretation of cystic fibrosis mutation analysis in clinical practice, *Journal of Cystic Fibrosis,* 7 (3), pp. 179–196.

Chapple, Alison, Champion, Peter, and May, Carl (1997) Clinical terminology: Anxiety and confusion amongst families undergoing genetic counseling, *Patient Education and Counseling,* 32, pp. 81–91.

Chargaff, Erwin (2001) Die wollen ewiges Leben, die wollen den Tod besiegen—das ist teuflisch, in *stern,* pp. 47244–252.

Chen, S., and Parmigiani, G. (2007) Meta-analysis of BRCA1 and BRCA2 penetrance, *J Clin Oncol,* 25 (11), pp. 1329–33.

Daston, Lorraine (1988) *Classical Probability in the Enlightenment,* Princeton, NJ: Princeton University Press.

Daston, Lorraine (1998) Die Kultur der wissenschaftlichen Objektivität, in Oexle, Otto Gerhard (ed.) *Naturwissenschaft, Geisteswissenschaft, Kulturwissenschaft: Einheit – Gegensatz – Komplementarität?,* Göttingen: Wallstein Verlag, pp. 9–39.

Daston, Lorraine (2000) The historicity of science, in Daston, Lorraine (ed.) *Biographies of Scientific Objects*, Chicago, London: University of Chicago Press, pp. 201–221.

Daston, Lorraine (2001) History of science, in Smelser, Neil, and Baltes, Paul (eds.) *International Encyclopedia of the Social and Behavioral Sciences*, Amsterdam: Elsevier, pp. 6842–6848.

Daston, Lorraine, and Galison, Peter (2007) *Objectivity*, New York: Zone Books.

Davison, C., Frankel, S., and Davey Smith, G. (1992) The limits of lifestyle—re-assessing fatalism in the popular culture of illness prevention, *Social Science and Medicine*, 34, pp. 675–685.

Dean, Mitchell (1998) Risk, calculable and incalculable, *Soziale Welt*, 49, pp. 25–42.

Degener, Theresia (1992) Humangenetische Beratung, pränatale Diagnose und (bundes)deutsche Rechtsprechung, in Stein, Anne-Dore (ed.) *Lebensqualität anstatt Qualitätskontrolle menschlichen Lebens*, Berlin: Edition Marhold, pp. 186–198.

Dijck, Jose v. (1998) *Imagenation: Popular Images of Genetics*, London: Macmillan.

Duden, Barbara (1991) *Der Frauenleib als öffentlicher Ort. Vom Mißbrauch des Begriffs Leben*, Hamburg: Luchterhand.

Duden, Barbara (1996) 'Das Leben' als Entkörperung. Anmerkungen zum Urteilsspruch zu Paragraph 218 durch das Bundesverfassungsgericht in Karlsruhe, Mai 1993, in Frauen gegen Bevölkerungspolitik (ed.) *LebensBilder – LebensLügen. Leben und Sterben im Zeitalter der Biomedizin*, Hamburg: Verlag Libertäre Assoziation, pp. 89–100.

Duden, Barbara (2000) Hoffnung, Ahnung, 'sicheres' Wissen. Zur Historisierung des Wissensgrundes vom Schwangergehen, *Die Psychotherapeutin*, 13, pp. 25–37.

Duden, Barbara (2002a) *Die Gene im Kopf – der Fötus im Bauch*, Hannover: Offizin.

Duden, Barbara (2002b) Zwischen 'wahrem Wissen' und Prophetie. Konzeptionen des Ungeborenen, in Schlumbohm, Jürgen, Veit, Patrice, and Duden, Barbara (eds.) *Geschichte des Ungeborenen. Zur Erfahrungs- und Wissenschaftsgeschichte der Schwangerschaft,*

17.-20, *Jahrhundert,* Göttingen: Vandenhoek & Ruprecht, pp. 11-48.

Duden, Barbara, and Samerski, Silja (2007) "Pop-genes": an investigation of "the gene" in popular parlance, in Burri, Regula Valérie, and Dumit, Joseph (eds.) *Biomedicine As Culture: Instrumental Practices, Technoscientific Knowledge, and New Modes of Life,* New York, London: Routledge, pp. 167-189.

Eiben, Bernd, and Glaubitz, Ralf (2005) Pränataldiagnostik — Erwiderung, *Deutsches Ärzteblatt,* 102 (36), pp. A2391-A2392.

Engeln, Henning (2006), Die Macht des Erbes, *GEOkompakt,* 7, pp. 8-19.

En Route to the Knowledge-Based Bio-Economy (2007) Conference Report, European Commission, Cologne, [Online] https://www.bmbf.de/pub/cp.pdf [8.8.2014].

Epigenome NoE (2008) *The Enemy Within,* [Online] http://epigenome.eu/en/2,54,0 [6.8.2014].

Erikson, Paul, Klein, Judy L., Daston, Lorraine, Lemov, Rebecca, Sturm, Thomas, and Gordin, Michael D. (2013) *How Reason Almost Lost Its Mind: The Strange Career of Cold War Rationality,* Chicago: University of Chicago Press.

Euhus, David (2001) Understanding mathematical models for breast cancer risk assessment and counselling, *The Breast Journal,* 7 (4), pp. 224-233.

Ewald, François (1991) Insurance and risk, in Burchell, Graham, Gordon, Colin, and Miller, Peter (eds.) *The Foucault Effect: Studies in Governmentality,* Chicago: University of Chicago Press, pp. 197-210.

Ewald, François (1993) *Der Vorsorgestaat,* Frankfurt a.M.: Suhrkamp (in the original: L'Etat providence, Paris: Editions Gasset & Fasquelle, 1986).

Falk, Raphael (1984) The gene in search of an identity, *Human Genetics,* 68, pp. 195-204.

Fasching, P.A., et al. (2007) Evaluation of mathematical models for breast cancer risk assessment in routine clinical use, *European Journal of Cancer Prevention,* 16 (3), pp. 216-224.

Fleck, Ludwik (1981) *Genesis and Development of a Scientific Fact,* Cambridge: Cambridge University Press.

Fleck, Ludwik (1986) Scientific observation and perception in general, in Cohen, R.S., and Schnelle, Thomas (eds.) *Cognition and Fact: Materials on Ludwik Fleck, Boston Studies in the Philosophy of Science, v. 87*, Dordrecht: D. Reidel, pp. 59–78.

Foucault, Michel (1990a) *The History of Sexuality: An Introduction*, New York: Vintage books.

Foucault, Michel (1990b) *The History of Sexuality: The Use of Pleasure*, New York: Vintage books.

Foucault, Michel (1982) The subject and power, *Critical Inquiry*, 8 (4), pp. 777–795.

Franklin, Sarah (1997) 'Having to try' and 'having to choose': How IVF 'makes sense', in Franklin, Sarah (ed.) *Embodied Progress: A Cultural Account of Assisted Conception*, New York: Routledge, pp. 168–245.

Fraser, F.C. (2001) Resetting our educational sights: Unconstructing the public's dreams and nightmares of the genetic revolution, *American Journal of Human Genetics*, 68, pp. 828–830.

Friedrich, Hannes, Henze, Karl H., and Stemann-Acheampong, Susanne (1998) *Eine unmögliche Entscheidung. Pränataldiagnostik: Ihre psychosozialen Voraussetzungen und Folgen*, Berlin: VWB.

Fuhrmann, Walter, and Vogel, Friedrich (1975) *Genetische Familienberatung: Ein Leitfaden für Studenten und Ärzte*, Berlin, Heidelberg, New York: Springer.

Galton, Francis (1883) *Inquiries into Human Faculty and its Development*, London: Macmillan.

Gemeinsamer Bundesausschuss (2007) *Beratungspflicht statt verpflichtende Früherkennungsuntersuchungen (Presseerklärung)*, [Online], http://www.g-ba.de/downloads/34-215-191/2007-07-20-Chroniker.pdf [6.4.2010].

GEOkompakt (2006) *Der Mensch und seine Gene*, Nr. 7.

Gibbon, Sahra, Kampriani, Eirini, and zur Nieden, Andrea (2010) BRCA patients in Cuba, Greece and Germany: Comparative perspectives on public health, the state and the partial reproduction of 'neoliberal' subjects, *BioSocieties*, 5 (4), pp. 440–466.

Gifford, S. (1986) The meaning of lumps; A case study of the ambiguities of risk, in Janes, C.R., Stall, R., and Gifford, S.M.

(eds.) *Anthropology and Epidemiology: Interdisciplinary Approaches to the Study of Health and Disease*, Dodrecht: Reidel, pp. 213–246.

Gigerenzer, Gerd (2003) *Reckoning with Risks: Learning to Live with Uncertainty*, New York: Penguin Books.

Gigerenzer, Gerd (2014) *Risk Savvy: How to Make Good Decisions*, London: Allen Lane.

Gigerenzer, Gerd, et al. (1989) *The Empire of Chance: How Probability Changed Science and Everyday Life*, New York: Cambridge University Press.

Gläsernes Labor (n.d. a) *Flyer*, [Online] http://www.berlin-buch-gesundheitsregion.de/download/FlyerGlaesernesLabor.pdf [3.4.2010].

Gläsernes Labor (n.d. b) *Erbsubstanz zum Anfassen I: Isolierung von DNA aus der Tomate*, [Online] http://www.glaesernes-labor.de/versuch1.html [10.4.2010].

Gøtzsche, Peter C. (2003) Mortality reduction by breast-cancer screening, *The Lancet*, 362 (9379), pp. 245–6.

Gøtzsche, Peter C., and Olsen, Ole (2000) is screening for breast cancer with mammography justifiable?, *The Lancet*, 355, pp. 129–134.

Greco, Monica (1993) psychosomatic subjects and the "duty to be well": personal agency within medical rationality, *Economy and Society*, 22 (3), pp. 355–372.

Grießler, Erich, Littig, Beate, and Pichelstorfer, Anna (2009) "Selbstbestimmung" in der genetischen Beratung: Argumentationsstruktur und Ergebnisse einer Serie neosokratischer Dialoge in Österreich und Deutschland, in Hirschberg, Irene, Grießler, Erich, Littig, Beate, and Frewer, Andreas (eds.) *Ethische Fragen genetischer Beratung. Klinische Erfahungen, Forschungsstudien und soziale Perspektiven*, Frankfurt am Main: Peter Lang, pp. 171–187.

Gronemeyer, Marianne (1988) *Die Macht der Bedürfnisse. Reflexionen über ein Phantom*, Reinbek bei Hamburg: Rowohlt.

Habermas, Jürgen (2001) *Die Zukunft der menschlichen Natur. Auf dem Weg zu einer liberalen Eugenik?*, Frankfurt am Main: Suhrkamp.

Hacking, Ian (1990) *The Taming of Chance,* Cambridge: Cambridge University Press.

Hallowell, N., et al. (2004) Accommodating risk: Responses to BRCA1/2 genetic testing of women who have had cancer, *Social Science and Medicine,* 59 (3), pp. 553–565.

Hallowell, N., and Richards, M.P.M. (1997) understanding life's lottery: an evaluation of studies of genetic risk awareness, *Journal of Health Psychology,* 2 (1), pp. 31–43.

Hallowell, Nina (1999) Doing the right thing: Genetic risk and responsibility, *Sociology of Health and Illness,* 21, pp. 597–621.

Hamer, Dean (2006) *Das Gottes-Gen. Warum uns der Glaube im Blut liegt,* München: Kösel.

Hammond, John S., Keeney, Ralph L., and Raiffa, Howard (1999) *Smart Choices: A Practical Guide to Making Better Decisions,* Boston, MA: Harvard Business School Press.

Heath, Deborah, Rapp, Rayna, and Taussig, Karen-Sue (2004) "Genetic citizenship", in Nugent, Frank A., and Vincent, Joan (eds.) *A Companion to the Anthropology of Politics,* Malden, Oxford: Blackwell Publishing, pp. 152–167.

Hermann, Svea L. (2009) *Policy Debates on Reprogenetics: The Problematization of New Research in Great Britain and Germany,* Frankfurt am Main: Campus.

Hirschberg, Irene, and Frewer, Andreas (2009) Genetik, Beratung und Ethik. Zur Einführung, in Hirschberg, Irene, Grießler, Erich, Littig, Beate, and Frewer, Andreas (eds.) *Ethische Fragen Genetischer Beratung. Klinische Erfahrungen, Forschungsstudien und soziale Perspektiven,* Frankfurt a.M.: Peter Lang, pp. 11–21.

Hubbard, Ruth, and Lewontin, Richard C. (1996) Pitfalls of genetic testing, *New England Journal of Medicine,* 334 (18), pp. 1192–1194.

Illich, Ivan (1992) Needs, in Sachs, Wolfgang (ed.) *Development Dictionary: A Guide to Knowledge as Power,* London, New Jersey: Zed Books, pp. 88–101.

Irwin, Alan, and Wynne, Brian (1996) Introduction, in Irwin, Alan, and Wynne, Brian (eds.) *Misunderstanding Science: The Public Reconstruction of Science and Technology,* Cambridge: Cambridge University Press, pp. 1–18.

Jacobsen, Hans-Jörg (2001) *Angemerkt: Prof. Dr. Hans-Jörg Jacobsen*, [Online] http://www.loccum.de/folo/angemerkt/jacobsen. html [8.1.2015].

Jansen, Sarah (2002) Den Heringen einen Paß ausstellen. Formalisierung und Genauigkeit in den Anfängen der Populationsökologie um 1900, *Berichte zur Wissenschaftsgeschichte*, 25, pp. 153–169.

Jennings, Bruce (2004) Genetic literacy and citizenship: Possibilities for deliberative democratic policymaking in science and medicine, *The Good Society*, 13 (1), pp. 38–44.

Johannsen, Wilhelm (1913) *Elemente der exakten Erblichkeitslehre*, Jena: G. Fischer.

Kant, Immanuel (1784/1997) *What is Enlightenment? Modern History Sourcebook, Fordham University*, [Online] http://www.fordham.edu/halsall/mod/kant-whatis.asp [12.8.2014].

Kant, Immanuel (1999) *Critique of Pure Reason*, Cambridge: Cambridge University Press.

Kaufert, Patricia, and O'Neil, John (1993) Analysis of a dialogue on risks in childbirth: Clinicians, epidemiologists, and Inuit women, in Lock, M., and Lindenbaum, S. (eds.) *Knowledge, Power and Practice: the Anthropology of Medicine and Everyday Life*, Berkeley, CA: University of California Press, pp. 32–54.

Kavanagh, A.M., and Broom, D.H. (1998) Embodied risk: my body, myself?, *Social Science and Medicine*, 46, pp. 437–444.

Kay, Lily (2000) *Who Wrote the Book of Life? A History of the Genetic Code*, Stanford, CA: Stanford University Press.

Keller, Evelyn Fox (2000) *The Century of the Gene*, Cambridge, MA: Havard University Press.

Kerr, Anne (2004) *Genetics and Society: A Sociology of Disease*, London: Routledge.

Kevles, Daniel (1985) *In the Name of Eugenics: Genetics and the Use of Human Heredity*, Berkeley, CA: University of California Press.

Kitcher, Philip (1992) Genes, in Fox Keller, Evelyn, and Lloyd, Elisabeth A. (eds.) *Keywords in Evolutionary Biology*, Cambridge, MA: Harvard University Press, pp. 128–131.

Klein, Stefan, and Venter, Craig (2009) Wir waren sehr naiv, *Zeitmagazin*, Nr. 03, 8.1.2009.

Koch, Lene, and Svendsen, Mette N. (2005) providing solutions, defining problems: The imperative of disease prevention in cancer genetic counselling, *Social Science and Medicine*, 60 (2005), pp. 823-832.

Kollek, Regine, and Lemke, Thomas (2008) *Der medizinische Blick in die Zukunft. Gesellschaftliche Implikationen prädiktiver Gentests*, Frankfurt a.M.: Campus.

Kommission für Öffentlichkeitsarbeit und ethische Fragen der Gesellschaft für Humangenetik e.V. (1996) Positionspapier, *Medizinische Genetik*, 8, pp. 125-131.

Kovác, Lázló, and Frewer, Andreas (2009) Die Macht medizinischer Metaphern: Studien zur Bildersprache in der genetischen Beratung und ihren ethischen Implikationen, in Hirschberg, Irene, Grießler, Erich, Littig, Beate, and Frewer, Andreas (eds.) *Ethische Fragen genetischer Beratung. Klinische Erfahungen, Forschungsstudien und soziale Perspektiven*, Frankfurt am Main: Peter Lang, pp. 205-221.

Krüger, Lorenz, Daston, Lorraine, and Heidelberger, Michael (1987) *The Probabilistic Revolution, vol. 1: Ideas in History*, Cambridge, MA: MIT Press.

Krüger, Lorenz, Gigerenzer, Gerd, and Morgan, Mary S. (1987) *The Probabilistic Revolution, vol. 2: Ideas in the Sciences*, Cambridge, MA: MIT Press.

Kuhlmann, Ellen (2002) *Ethik von Screening Rationalitäten – Teil I: Methodische Probleme und ungelöste Wertkonflikte*, Endbericht Hamburg. (Gefördert durch das BMBF, Förderkennzeichen 01KU9907).

Lemke, Thomas (2000) Die Regierung der Risiken. Von der Eugenik zur genetischen Gouvernmentalität, in Bröckling, Ulrich, Krasmann, Thomas, and Lemke, Thomas (eds.) *Gouvernmentalität der Gegenwart. Studien zur Ökonomisierung Des Sozialen*, Frankfurt am Main: Suhrkamp, pp. 227-264.

Lemke, Thomas (2004) *Veranlagung und Verantwortung. Genetische Diagnostik zwischen Selbstbestimmung und Schicksal*, Bielefeld: transcript.

Lemke, Thomas (2005) From eugenics to the government of genetic risk, in Bunton, Robin, and Petersen, Alan (eds.) *Genetic*

Governance: Health, Risk and Ethics in the Biotech Era, London, New York: Routledge, pp. 95–105.

Lenz, Widukind, and Lenz, Fritz (1968) Grundlinien der Humangenetik. Zu Definition, Terminologie und Methoden, in Becker, Peter Emil (ed.) *Humangenetik. Ein kurzes Handbuch in fünf Bänden*, Stuttgart: Thieme, pp. 1–76.

Lewontin, Richard (1992) Genotype and phenotype, in Lloyd, Elisabeth A., and Keller, Evelyn Fox (eds.) *Keywords in Evolutionary Biology*, Cambridge, MA: Havard University Press, pp. 137–144.

Lewontin, Richard (2004) The genotype/phenotype distinction, *The Stanford Encyclopedia of Philosophy*, [Online] http://plato.stanford.edu/entries/genotype-phenotype/ [12.2.2010].

Lippman, Abby (1991) Prenatal genetic testing and screening: Constructing needs and reinforcing inequities, in Clarke, Angus (ed.) *Genetic Counseling: Practices and Principles*, London, New York: Routledge, pp. 142–186.

Lippman, Abby (1994) The genetic construction of prenatal testing: Choice, consent or conformity for women?, in Rothenberg, Karen H., and Thompson, Elizabeth J. (eds.) *Women and Prenatal Testing: Facing the Challenges of Genetic Technology*, Columbus, OH: Ohio State University Press, pp. 9–34.

Lippman-Hand, Abby, and Fraser, F. Clarke (1979a) Genetic counseling: Provision and reception of information, *American Journal of Medical Genetics*, 3, pp. 113–127.

Lippman-Hand, Abby, and Fraser, F. Clarke (1979b) Genetic counseling: The postcounseling period: I. Parent's perception of uncertainty, *American Journal of Medical Genetics*, 4, pp. 51–71.

Lippman-Hand, Abby, and Fraser, F. Clarke (1979c) Genetic counseling: The postcounseling period: II. Making reproductive choices, *American Journal of Medical Genetics*, 4, pp. 73–87.

Lock, Margaret (1998) Breast cancer: Reading the omens, *Anthropology Today*, 14 (4), pp. 7–16.

Lock, Margaret (2009) Testing for susceptibility genes: A cautionary tale, in Rehmann-Sutter, Christoph, and Müller, Hansjakob (eds.) *Disclosure Dilemmas: Ethics of Genetic Prognosis After the*

'Right to Know/Not to Know' Debate, Farnham, Burlington: Ashgate, pp. 65-84.

Lock, Margaret, et al. (2006) When it runs in the family: Putting susceptibility genes into perspective, *Public Understanding of Science*, 15, pp. 277-300.

Lux, M.P., Fasching, P.A., and Beckmann, M.W. (2006) Hereditary breast and ovarian cancer: Review and future perspectives, *Journal of Molecular Medicine*, 84 (1), pp. 16-28.

M'charek, Amâde (2000) Technologies of population: Forensic DNA testing practices and the making of differences and similarities, *Configurations*, 8 (1), pp. 121-158.

Martin, Emily (1987) *The Woman in the Body: A Cultural Analysis of Reproduction*, Boston, MA: Beacon Press.

Martin, Emily (1994) *Flexible Bodies: Tracking Immunity in American Culture – From the Days of Polio to the Age of AIDS*, Boston, MA: Beacon Press.

Maxton-Küchenmeister, Jörg, and Dähnhardt, Dorothee (2005) *Genlabor & Schule. Dokumentation eines Schülerlabor-Netzwerks*, Frankfurt und Kiel, [Online] http://www.genlabor-schule.de/download/Dokumentation:Genlabor_Schule.pdf [30.5.2010].

May, Stefan, and Holzinger, Markus (2003) *Autonomiekonflikte der Humangenetik. Professionssoziologische und professionsrechtliche Aspekte einer Theorie reflexiver Modernisierung*, Opladen: Leske und Budrich.

Minwegen, Norbert (2003) *Biotechnologie und Gentechnik zum Anfassen*, [Online] http://www.nrwchemie.de/news/archiv/010328_2htm [30.5.2010].

Mirowski, Philip (2002) *Machine Dreams: How Economics Became a Cyborg Science*, Cambridge: Cambridge University Press.

Mühlhauser, Ingrid (2013) Screening auf Brustkrebs/Mammografie-Screening, *Deutsche Zeitschrift für Onkologie*, 45, pp. 80-85.

Müller-Jung, Joachim (2006) Am Gen erstickt, *FAZ*, 20.9.2006, N1.

Nationales Netzwerk Frauengesundheit (2007) *Brustkrebs Früherkennung. Informationen zur Mammografie – eine Entscheidungshilfe. Aktualisierte Auflage*, Hamburg: Techniker Krankenkasse, [Online] http://www.tk-online.de/centaurus/

servlet/contentblob/92922/Datei/2572/TK-Broschuere-Brustkrebs-Frueherkennung.pdf [4.4.2010].

zur Nieden, Andrea (2013) *Zum Subjekt der Gene werden. Subjektivierungsweisen im Zeichen der Genetisierung von Brustkrebs*, Bielefeld: transcript.

Niewöhner, Jörg (2004) Integration von Bürgerbeteiligung als politische Aufgabe?, in Tannert, Christof, and Wiedemann, Peter (eds.) *Stammzellen im Diskurs. Ein Lese- und Arbeitsbuch zu einer Bürgerkonferenz*, München: Oekom, pp. 67-74.

Nikolow, Sybilla, and Schirrmacher, Arne (2007) *Wissenschaft und Öffentlichkeit als Ressource füreinander. Studien zur Wissenschaftsgeschichte im 20. Jahrhundert*, Frankfurt a.M.: Campus.

Nippert, Irmgard (2001) *Was kann aus der bisherigen Entwicklung der Pränataldiagnostik für die Entwicklung von Qualitätsstandards für die Einführung neuer Verfahren wie der Präimplantationsdiagnostik gelernt werden? Fortpflanzungsmedizin in Deutschland. Schriftenreihe des Bundesministeriums für Gesundheit*, Bonn: Nomos, pp. 293-321.

Noschka-Roos, Annette, and Teichmann, Jürgen (2006) Populäre Wissenschaft in Museen und Science Center, in Faulstich, Peter (ed.) *Öffentliche Wissenschaft. Neue Perspektiven in der Wissenschaftlichen Weiterbildung*, Bielefeld: transcript, pp. 87-103.

Novas, Carlos, and Rose, Nikolas (2000) Genetic risk and the birth of the somatic individual, *Economy and Society*, 29 (4), pp. 485-513.

Oakley, Ann (1984) *The Captured Womb: A History of Medical Care of Pregnant Women*, Oxford: Basil Blackwell.

Pap, Michael (1995) Genetische Beratung und Nichtdirektivität im Licht der zivilrechtlichen Haftungsrechtsprechung, in Ratz, Erhard (ed.) *Zwischen Neutralität und Weisung. Zur Theorie und Praxis von Beratung in der Humangenetik*, München: Evangelischer Presseverband für Bayern, pp. 51-56.

Paul, Diane B. (1995) *Controlling Human Heredity: 1865 to the Present*, New Jersey: Humanities Press.

Petersen, Alan (2002) Facilitating autonomy: The discourse of genetic counselling, in Peterson, Alan, and Bunton, Robin (eds.)

The New Genetics and the Public's Health, London: Routledge, pp. 135–158.

Petersen, Alan, and Bunton, Robin (2002) *The New Genetics and the Public's Health*, London: Routledge.

Pijpe, A., et al. (2012) Exposure to diagnostic radiation and risk of breast cancer among carriers of BRCA1/2 mutations: retrospective cohort study (GENE-RAD-RISK), *British Medical Journal*, 345, p. e5660.

Pilnick, Alison (2002) What "most people" do: Exploring the ethical implications of genetic counselling, *New Genetics and Society*, 21 (3), pp. 339–350.

Pörksen, Uwe (1986) *Deutsche Naturwissenschaftssprachen: Historische und kritische Studien*, Tübingen: Narr.

Pörksen, Uwe (1995) *Plastic Words: The Tyranny of a Modular Language*, State College, PA: Penn State University Press.

Pörksen, Uwe (1997) *Weltmarkt der Bilder. Eine Philosophie der Visiotype*, Stuttgart: Klett-Cotta.

Pörksen, Uwe (2003) Das Phlogiston — ein Wissenschaftsmärchen, in Hiß, Christian (ed.) *Der GENaue Blick. Grüne Gentechnik auf dem Prüfstand*, München: ökom-Verlag, pp. 76–81.

Porz, Rouven (2009) The need for an ethics of kinship: Decision stories and patients' context, in Rehmann-Sutter, Christoph, and Müller, Hansjakob (eds.) *Disclosure Dilemmas: Ethics of Genetic Prognosis After the 'Right to Know/Not to Know' Debate*, Farnham, Burlington: Ashgate, pp. 53–64.

Rapp, Rayna (1999) *Testing Women, Testing the Fetus: The Social Impact of Amniocentesis in America*, New York: Routledge.

Reed, Sheldon (1974) A short history of genetic counseling, *Social Biology*, 21 (4), pp. 332–339.

Reif, Maria, and Baitsch, Helmut (1986) *Genetische Beratung. Hilfestellung für eine selbstverantwortliche Entscheidung?*, Berlin: Springer.

Rheinberger, Hans-Jörg, and Müller-Wille, Staffan (2009) Gene, *Stanford Encyclopedia of Philosophy*, [Online] http://plato.stanford.edu/entries/gene/ [26.3.2010].

Rheinberger, Hans-Jörg, and Müller-Wille, Staffan (2009) *Vererbung. Geschichte und Kultur eines biologischen Konzeptes*, Frankfurt am Main: Fischer Taschenbuch Verlag.

Rhodes, Rosamond (1998) Genetic links, family ties, and social bonds: Rights and responsibilities in the face of genetic knowledge, *Journal of Medicine and Philosophy*, 23 (1), pp. 10–30.

Rieger, Matthias (2010) Lo ›sguardo statistico‹: l'adattamento dell'occhio umano alla società della sorveglianza, *Studi sulla questione criminale*, 5, pp. 101–112.

Robins, Rosemary (2001) Overburdening risk: Policy frameworks and the public uptake of gene technology, *Public Understanding of Science*, 10 (1), pp. 19–36.

Rose, Nikolas (1999) *Powers of Freedom: Reframing Political Thought*, Cambridge: Cambridge University Press.

Rose, Nikolas (2007) *The Politics of Life Itself: Biomedicine, Power, and Subjectivity in the Twenty-First Century*, Princeton, NJ: Princeton University Press.

Rothman, Barbara K. (1986) *The Tentative Pregnancy: Prenatal Diagnosis and the Future of Motherhood*, New York: Penguin.

Rothman, Barbara K. (1998) *Genetic Maps and Human Imaginations: The Limits of Science in Understanding Who We Are*, New York: W.W. Norton & Company.

Rothman, David (1991) *Strangers at the Bedside: A History How Law and Bioethics Transformed Medical Decision Making*, New York: Basic Books.

Samerski, Silja (2002) *Die verrechnete Hoffnung. Von der selbstbestimmten Entscheidung durch genetische Beratung*, Münster: Westfälisches Dampfboot.

Samerski, Silja (2009) Genetic counseling and the fiction of choice: Taught self-determination as a new technique of social engineering, *Signs: Journal of Women in Culture and Society*, 34, pp. 735–761.

Samerski, Silja (2015) Pregnancy, personhood and the making of the fetus, in Disch, Lisa, and Hawkesworth, Mary (eds.) *Oxford Handbook of Feminist Theory*, Oxford: Oxford University Press.

Sarangi, Srikant (2002) The language of likelihood in genetic-counseling discourse, *Journal of Language and Social Psychology*, 21, pp. 7–31.

Schicktanz, Silke, and Naumann, Jörg (2003) *Bürgerkonferenz: Streitfall Gendiagostik. Ein Modellprojekt der Bürgerbeteiligung am bioethsichen Diskurs*, Opladen: Leske + Budrich.

Schimank, Uwe (2005) *Die Entscheidungsgesellschaft. Komplexität und Rationalität der Moderne*, Wiesbaden: VS, Verlag für Sozialwissenschaften.

Schirmer, Christine (2009) Genetische Beratung aus Betroffenenperspektive. Von der Diagnose zur Entscheidung – ein Erfahrungsbericht, in Hirschberg, Irene, Grießler, Erich, Littig, Beate, and Frewer, Andreas (eds.) *Ethische Fragen genetischer Beratung. Klinische Erfahrungen, Forschungsstudien und Soziale Perspektiven*, Frankfurt am Main: Peter Lang, pp. 121–128.

Schmidtke, Jörg (1995) Die Indikationen zur Pränataldiagnostik müssen neu begründet werden, *Medizinische Genetik*, 1, pp. 49–52.

Schmidtke, Jörg (2008) Gentests: Auf dem Prüfstand der Genetiker, *Deutsches Ärzteblatt*, 105 (36), pp. A 1830–A 1834.

Schmidtke, Jörg, and Pabst, Brigitte (2007) Daten zu ausgewählten Indikatoren II, in Schmidtke, Jörg, and Röber, Bernd (eds.) *Gendiagnostik in Deutschland. Status Quo und Problemerkundung. Supplement Zum Gentechnologiebericht*, Berlin: Berlin-Brandenburgische Akdademie der Wissenschaften, pp. 195–203.

Schmidtke, Jörg, and Rüping, Uta (2013) Genetische Beratung: Nichtärztliche Personen können ein Gewinn sein, *Deutsches Ärzteblatt*, 110 (25), pp. A1248–50.

Schmutzler, R., et al. (2003) Beratung, Genetische Testung und Prävention von Frauen mit einer familiären Belastung für das Mamma- und Ovarialkarzinom, *Zentralblatt für Gynäkologie*, 125 (12), pp. 494–506.

Schwartz, Sara (2000) The differential concept of the gene, past and present, in Beurton, Peter J., Falk, Raphael, and Rheinberger, Hans-Jörg (eds.) *The Concept of the Gene in Development and Evolution: Historical and Epistemological Perspectives*, Cambridge: Cambridge University Press, pp. 26–39.

Schwarz, Clarissa M., and Schücking, Beate (2004) Adieu, normale Geburt? Ergebnisse eines Forschungsprojektes, *Dr.med.Mabuse*, 148 (March/April), pp. 22-25.

Schwennesen, Nete, Koch, Lene, and Svendsen, Mette N. (2009) Practising informed choice: Decision making and prenatal risk assessment—the Danish experience, in Rehmann-Sutter, Christoph, and Müller, Hansjakob (eds.) *Disclosure Dilemmas: Ethics of Genetic Prognosis After the 'Right to Know/Not to Know' Debate*, Farnham, Burlington: Ashgate, pp. 191-204.

Selting, Margret (1998) Gesprächsanalytisches Transkriptionssystem (GAT), *Linguistische Berichte*, 173, pp. 91-122.

Sen, Amartya K. (1977) Rational fools: A critique of the behavioural foundation of economic theory, *Philosophy & Public Affairs*, 6 (4), pp. 317-344.

Shapin, Steven (1994) *A Social History of Truth: Civility and Science in Seventeenth-Century England*, Chicago, London: University of Chicago Press.

Shea, Elizabeth (2001) The gene as a rhetorical figure: "nothing but a very applicable little word", *Science as Culture*, 10 (4), pp. 505-530.

Slieker, M.G., et al. (2005) Disease modifying genes in cystic fibrosis, *Journal of Cystic Fibrosis*, 4 (suppl 2), pp. 7-13.

Soden, Kristine v. (1988) *Die Sexualberatungsstellen der Weimarer Republik 1919-1933*, Berlin: Edition Hentrich.

Steiner, Patricia, Gadzicki, Dorothea, and Schlegelberger, Brigitte (2009) Probleme bei der Weitergabe der genetischen Information innerhalb von Familien mit erblichem Brust- und Eierstockkrebs—nur eine Familienangelegenheit?, in Hirschberg, Irene, Grießler, Erich, Littig, Beate, and Frewer, Andreas (eds.) *Ethische Fragen Genetischer Beratung. Klinische Erfahrung, Forschungsstudien und Soziale Perspektiven*, Frankfurt am Main: Peter Lang, pp. 25-50.

Strathern, Marilyn (1995) Displacing knowledge: Technology and the consequence for kinship, in Ginsburg, Faye D., and Rapp, Rayna (eds.) *Conceiving the New World Order: The Global Politics of Reproduction*, Berkeley, CA: University of California Press, pp. 346-363.

Strohman, Richard C. (1997) The coming Kuhnian revolution in biology, *Nature Biotechnology*, 15 (3), pp. 194–200.
Swiss Medical Board (2013) *Systematisches Mammographie-Screening. Bericht vom 15. Dezember 2013*, [Online] http://www.medical-board.ch/fileadmin/docs/public/mb/Fachberichte/2013-12-15_Bericht_Mammographie_Final_rev.pdf [13.8.2014].
Tannert, Christof, and Wiedemann, Peter (2004) *Stammzellen im Diskurs. Ein Lese- und Arbeitsbuch zu einer Bürgerkonferenz*, München: Oekom.
The Cystic Fibrosis Genotype-Phenotype Consortium (1993) Correlation between genotype and phenotype in patients with cystic fibrosis, *The New England Journal of Medicine*, 329 (18), pp. 1308–1313.
Theile, Ursel (1977) *Genetische Beratung: Motivationsanalyse*, München: Urban & Schwarzenberg.
Turner, John R.G. (2001) Self-made men (Review), *Times Literary Supplement*, 5142, p. 8.
Waldschmidt, Anne (1996) *Das Subjekt in der Humangenetik. Expertendiskurse zur Programmatik und Konzeption der genetischen Beratung 1945–1990*, Münster: Westfälisches Dampfboot.
Watkins, Elizabeth S. (1998) *On the Pill: A Social History of Oral Contraceptives 1950–1970*, Baltimore, MD: Johns Hopkins University Press.
Weingart, Peter, Kroll, Jürgen, and Bayertz, Kurt (1992) *Rasse, Blut und Gene. Geschichte der Eugenik und Rassenhygiene in Deutschland*, Frankfurt am Main: Suhrkamp.
Weir, Lorna (1996) Recent developments in the government of pregnancy, *Economy and Society*, 25 (3), pp. 372–392.
Weir, Lorna (2006) *Pregnancy, Risk, and Biopolitics: On the Threshold of the Living Subject*, London: Routledge.
Wendt, G. Gerhard (1978) Die Notwendigkeit der genetischen Beratung, *Internist*, 19, pp. 441–444.
Wertz, Dorothy C., and Fletcher, John C. (2004) *Genetics and Ethics in Global Perspective*, Dordrecht: Kluwer Academic Publishers.
Weß, Ludger (1989) *Die Träume der Genetik. Gentechnische Utopien von sozialem Fortschritt*, Nördlingen: Greno.

Weymayr, Christian, and Koch, Klaus (2003) *Mythos Krebsvorsorge. Schaden und Nutzen der Früherkennung*, Frankfurt a.M.: Eichborn.

Wiener, Norbert (1950) *The Human Use of Human Beings: Cybernetics and Society*, Boston, MA: Houghton Mifflin.

Wynne, Brian (1995) Public understanding of science. in Jasanoff, Sheila, Markle, Gerald E., Petersen, James C., and Pinch, Trevor (eds.) *Handbook of Science and Technology Studies*, Sage, pp. 361–388.

Wynne, Brian (1996) Misunderstood misunderstandings: Social identites and public uptake of science, in Irwin, Alan, and Wynne, Brian (eds.) *Misunderstanding Science? The Public Reconstruction of Science and Technology*, Cambridge: Cambridge University Press, pp. 19–46.

Zoll, Barbara (2009) Autonomie, Entscheidungsfindung und Nicht-Direktivität in der genetischen Beratung—eine ethische Betrachtung, in Hirschberg, Irene, Grießler, Erich, Littig, Beate, and Frewer, Andreas (eds.) *Ethische Fragen genetischer Beratung. Klinische Erfahungen, Forschungsstudien und soziale Perspektiven*, Frankfurt am Main: Peter Lang, pp. 85–102.

Zuuren, F.J. v., Schie, E.C.M. v., and Baaren, N.K. v. (1997) Uncertainty in the information provided during genetic counseling, *Patient Education and Counseling*, 32, pp. 129–139.

Index

Abortion 38, 41, **42**, 125, 126
Amniocentesis 28, 38, 40, 43, 92, 93, **123-8**, 130-4, 137, 142
Autonomy (see also: self-determination) viii, ix, xi, 2, 5, 16, 21, 30, 48, 51, 106, 108, 109, 112, 114, 117, 122, 128, 140, 142 fn2, 143
Bioinformatics **27 fn11**
Blood test **40**
BRCA 89 fn33, 97, 98, 100
Breast Cancer 46, 72, 83 fn27, 88, 89 fn33, **96-102**, 112, 118, **119-22**, 143

Cause (in genetics and statistics) 5, 8, 11 fn4, 13 fn8, 14, **65-8**, 71, 72, **73**, 74, 75, 96, 110, 111
Colloquial speech/language xii, 3 fn2, **4**, 5, 24, 49 fn2, 56, 58, 86, 87, 128, 129
Colon cancer 46-9, 67, 68, 72, 78, 81-4, 87, 101-3, 109, 113, 114, 116, 121
Colonoscopy 116, 121

Correlation 5, **13**, **66**, 67, 73, 98, 124, 142
Cost-benefit 122, 128, **130-4**
Cybernetics (see also: immune system) 10, **69**, 70
Cystic fibrosis 6, 39 fn48, **65**, 66, 92, 99

Decision sciences, decision theory xii, **128**, **129**
Decision trap vii, viii, x, xii, 134, 135, **138**, **139**, 141, 143, 146
Down's syndrome (see also trisomy 21) 28 fn33, **38**, 40 fn50, 88 fn32, 90, 92, 93, 124, 131 fn63, 132, 133, 138 fn69
Early detection 29, 48, 99, 101, 111, 113, 115, 116, **121**, **122**
Epistemic transformation/ epistemic confusion **85-7**, 99, 105
Eugenics viii, 15, 17, **31-7**, 38, 41, 44, 45, 76 fn22, 126, 141

Gene (as a term or concept) x, xiii, 1, 2, 5, 7, **8–16**, 17, 25–7, 32, 47, 48, **49–77**, 93, 95, 96, 101, 102, **103–5**, 109, 110, 111 fn48, 112–4, 120, 146

"Gene for" 6, 20, 39 fn48

Genetic citizenship 21, 22

Genetic literacy 2, 15, **21**, **22**, 25 fn27, 34

Genotype-phenotype **13**, 39 fn48, 65 fn14, 66, 67, 72

Hypostatization (of the gene) 74–6

Immune system 102, 103

Information vii, viii, ix, 2, **3**, 16 fn17, 54, 56, 77, 82, 129, 130

Information, genetic 8, 10 fn3, 26, 51, 55, **68–72**, 74, 75, 77

Irrealis mood 89, **94**, **95**, 100, 101

Laws (judicial) 30, 35, 38, 41–3, 90, 107

Mammography 121, 122

Monitoring (see also: surveillance) 36 fn44, 40, 41, 83, 102, 103, 114, 116, 117 fn52, 120, 122 fn58, 123, 126

Normalcy 37, 38, 40, 41, 62, 63, 67, 80 fn25, 88, 91, 109, 110, 138 fn69

Normalization **15** fn14

Participation (democratic) 1, 2, **15–23**

Popularization, popular science 12 fn6, 16 fn17, **26**, 34, 56–61, 70, **73–7**

Pregnancy, pregnant women xi, xii, 5, 6, 28, 29, 36, **37–44**, 46, 54, 68, 69, 88 fn32, 89, **92–5**, 99, 108, **123–39**, 142 fn2, 145

Prenatal testing, prenatal diagnostics (see also: amniocentesis and blood test) viii, ix, 12, 29, 30, **38–43**, 44, 126, **131–9**, 142

Prevention 36, 37, 40, 44, 105, 112, 116 fn50, 118, 119, 143

Rational choice (see decision sciences and rationality, managerial)

Rationality, managerial or economic xii, **126–34**

Responsibility xi, 2–4, 16, 17, 19, 33, 34, 40, 42, 44, 48, 102, 107, 109, 112, 114–8, 136–40, 142 fn2, 143, 144

Risk (as a term, as a concept) **78–81**, **86**, **87**

Risk, diagnosis of and risk, personal **77–89**, 90–5, **96–102**, 103–6, 111–8, 120, 122, 124

Risk management 24, 36, 41, 115, 117–9, 126, 136, 137, 142–4

Risk profile 44, 79, 81-3, 85-7, 91, 94, 98-100, 109, 116, 123, 134-7, 139, 141-6

Scientific facts (see also: "gene") 4-7, **9-11**, 26, 34, 56, 58-61, 71, 75, 76

Self-determination (see also: autonomy) xi, 3, 16, 20, 30, 32, 48, 49, **51**, 100 fn43, 115, 117, 123, 140, 141, **144**

Self-management (see also: risk management) 102, 103, 108, 109, 119, 135, 136

Self-perception xii, 27, 51, 52, 63, 64, 68, 71, 72, 74, 77, 83, **102-5**, **144**, **145**

Social control (see also: social technology) viii, **3**, 34 fn40, 69 fn18, 144 fn3

Social technology (see also: social control) 40, 143, 144

Statistical construct(ion) 13, 15 fn14, 26 fn31, 79, 81, 82, 84, 87, 88, 91, 95, 99, 127, 132 fn65, 142, 143, 146

Surveillance (see also: monitoring) 104, 117, 122, 136 fn68

Surveillance medicine **40**, **41**, 116 fn51

Susceptibility (incl. genetic susceptibility) **13**, 44, 47, 80, 96, 100

Trisomy 21 (see also Down's syndrome) 38, 91-3, 133

CPSIA information can be obtained
at www.ICGtesting.com
Printed in the USA
BVOW06s0845111216
470457BV00021B/424/P